Fundamentals of Chemical Kinetics

Fundamentals of Chemical Kinetics

S. R. Logan

Reader in Chemistry
University of Ulster

Longman

Longman Group Limited
Longman House, Burnt Mill, Harlow
Essex CM20 2JE, England
and Associated Companies throughout the world

First published 1996

British Library Cataloguing in Publication Data
A catalogue entry for this title is available from the British Library.

ISBN 0-582-25185-0

Library of Congress Cataloging-in-Publication Data
A catalog entry for this title is available from the Library of Congress.

Typeset by 16 10/12 point Times Roman (Monotype)
Produced through Longman Malaysia, VVP

To my wife, Renée, and our sons, Bruce and Alan

Contents

Preface

When teaching chemical kinetics to successive cohorts of undergraduates I have often wished that there existed an inexpensive textbook possessing all the conceivable virtues. It should, as far as possible, treat all aspects of kinetics, rather than leave important topics unmentioned. It should do justice to both experiment and theory, putting the role of each in context. It should explain concepts in ways that students might be expected to understand, aiming for the merits of lucidity and simplicity rather than for excessive erudition.

In December 1993, I was invited to prepare my own undergraduate text on kinetics. In doing this I have started at the beginning since I believe that many difficulties about kinetics are derived from an incomplete understanding of the basics. Technological advances have resulted in the widespread use of many kinetic techniques, not merely by physical chemists, but by organic (and not simply physical organic) chemists, inorganic chemists and biochemists. A growing proportion of the ever-expanding chemical literature contains some kinetic component. For all such workers, the fundamentals of the subject are vitally important. To avoid too large a compendium, in regard to various aspects I have had to make my selection. In so doing, my central aim has been to illustrate the principles involved.

Regarding terminology and units, I have tried to adhere to the best practices which are conventional in the current literature. One point of departure is that I have restricted the use of an arrow ("→") to elementary processes. This I have done because, to understand the kinetics of composite reactions, it is essential that individual reaction steps be clearly distinguished from the overall reaction. The IUPAC recommendation of two different kinds of arrows seems much too subtle a distinction, so overall reactions have been written as balanced equations using an equality sign, "=".

In writing this book I have had much assistance from friends in Coleraine and elsewhere. John Stewart, formerly of UU Jordanstown and Gertie Taggart, now at Athlone RTC, both plowed through the first draft and marked my errors and omissions. The reviewer enlisted by my publishers made helpful and penetrating observations. Gerry Shannon of Mathematics solved many equations for me. Elaine Urquhart and the Library staff have fetched lots of volumes from obscure corners and from elsewhere by Inter-

Library loan. Dick Jones of Kent and my Coleraine colleagues Norman Brown and Chris Knipe each gave me detailed comment on one draft chapter. Numerous other colleagues have helped me in regard to specific points and I thank them all. Despite all this assistance and my avowed intention to avoid perpetuating errors, I have no doubt that some have filtered through and for these I must accept responsibility.

The first part of my manuscript was typed in the departmental office by Carol Millar and the second part plus all the revisions by my wife Renée. Both had other things to do and to both I am deeply indebted. For the publishers, Kathy Hick and Chris Leeding were most co-operative.

S. R. Logan
University of Ulster at Coleraine March 1995

Acknowledgements

We are grateful to the following for permission to reproduce copyright material: the Royal Society and Lord Porter for Table 3.1; the Royal Society of Chemistry for Figure 12.8 and Table 8.1; the American Chemical Society for Figure 9.6; the American Institute of Physics and Professor Y. T. Lee for Figure 12.3.

1 The empirical framework of chemical kinetics

Among the various branches of physical chemistry, such as chemical thermodynamics or spectroscopy, chemical kinetics is unusual in that it interprets and reports experimental data in terms of essentially empirical parameters. In this opening chapter, the concepts which make up this empirical framework are introduced and explained. Their use in the treatment of data on the progress of chemical reactions is then illustrated in a series of selected examples.

1.1 Introduction

One of the basic notions in chemistry is that of a chemical reaction, in which one substance (or group of substances) is converted into another. For such a process, which involves a rearrangement of the bonding electrons of the constituent atoms, a balanced or stoichiometric equation may be written, as in the following examples:

$$2O_3 = 3O_2 \tag{1.1}$$

$$CH_3CO_2C_2H_5 + H_2O = CH_3CO_2H + C_2H_5OH \tag{1.2}$$

$$BrO^-_3 + 5Br^- + 6H^+ = 3Br_2 + 3H_2O \tag{1.3}$$

Quantitative studies of chemical reactions tend to fall into one of two groups. There are those relating to the actual occurrence of the reaction, regardless of how quickly or how slowly it takes place. These may measure such quantities as the standard enthalpy change or the standard Gibbs energy change of the reaction. The latter may lead to a knowledge of the extent of reaction expected under specified conditions, for chemical equilibrium to be attained between the reacting substances, written on the left hand side of the equation and the product substances written on the right.

The second group of studies relates to the rapidity with which a chemical reaction occurs and, unlike the first group, it uses time as a variable. Also, it involves a clearly defined concept, called the *reaction rate*. Studies of the dependence of the rate on a range of factors can provide an empirical basis on which one might predict how rapidly the reaction will occur under hypothetical conditions.

In addition, they may provide relevant information about the detailed manner in which the chemical reaction takes place. In some instances they may demonstrate only that the reaction does not occur in a certain way. In others, they may suffice to show that the total chemical process represented in the stoichiometric equation comes about through a particular succession of events. Thus this second group of studies can potentially generate much information about a reaction that could never be discovered on the basis of even the most extended experiments of the first group.

The subject matter of this second group of studies is called *chemical kinetics*. The word "kinetics" (from the Greek, *kīnētikos* = moving) carries the connotation of movement or change. The term is used by chemists to denote not only studies of reaction rate and its dependence on the relevant parameters, but also an understanding of the factors which determine how quickly chemical reactions take place. Thus it encompasses experimental measurement as well as empirical and theoretical approaches to the interpretation of these measurements.

The inter-relationship of chemical thermodynamics and chemical kinetics is an issue of some interest and complexity. However, for chemical reactions that are readily amenable to kinetic study, the reactant and product species are usually all in the same homogeneous phase. Regardless of whether this is the gas phase or solution, thermodynamics tells us that only a certain extent of reaction is possible. For the system to progress beyond this extent of reaction would offend the Second Law of Thermodynamics. On the other hand, there are no laws stipulating how quickly this extent of reaction is approached. The reaction may well be so slow that this is unobservable.

Let us consider the homogeneous gas phase reaction:

$$2HI = H_2 + I_2 \tag{1.4}$$

Using tabulated data, the standard Gibbs energy change $\Delta G^\ominus$ of this reaction at 600 K may be estimated as 23.17 kJ mol^{-1}. The equilibrium constant, K_p, of this reaction may be expressed in terms of the degree of advancement, ξ, where $\xi = 0$ denotes no reaction and $\xi = 1$ represents complete conversion to the products. Thus we obtain the equation,

$$\Delta G^\ominus = -RT \ln\left\{\frac{\xi^2}{4(1-\xi)^2}\right\} \tag{1.5}$$

which leads to

$$\frac{\xi}{2(1-\xi)} = \exp\left(\frac{-\Delta G^\ominus}{2RT}\right) = 0.0981 \tag{1.6}$$

and thus to $\xi = 0.164$. This means we can deduce that at 600 K, reaction (1.4) will proceed only about one-sixth of the way to completion.

It is pertinent to remember that a homogeneous chemical reaction approaching equilibrium is a dynamic system, with continual interconversion of reactants and products. A highly positive value of the Gibbs energy change may well mean that only a very small degree of advancement of the reaction can be attained, but, no matter how small this is, this does not mean that the process of the forward reaction does not occur. Rather, the situation arises because at that minute value of ξ, the rates of the forward and the reverse reactions are equal.

In recent years, the respective roles of thermodynamics and kinetics may have been altered slightly. Where a chemical species is extremely reactive and inevitably undergoes reaction within a small fraction of a second, the usual means of measuring thermodynamic properties are not appropriate. In such a case, it *may* be possible to derive the standard Gibbs energy of formation of the unstable species by using kinetic data, and if the assumptions which require to be made are justified, then the value so obtained may be quite reliable. (A similar procedure is equally possible in the case of a stable species, but there is simply no need to resort to it.) This emphasises the complementary nature of these two areas of physical chemistry.

1.2 The rate equation

To express the rapidity of a chemical reaction in a quantitative manner requires a definition of the reaction rate. As this concept is normally used, it denotes the rate of change of concentration concomitant with reaction. Considering the hypothetical reaction,

$$\mathrm{A} = \mathrm{B} \tag{1.7}$$

taking place in a closed space, the instantaneous rate of this reaction is given by the rate of decrease of the concentration of A or by the rate of increase of the concentration of B.

To express the rate of change in a precise manner, it is necessary to use calculus. For reaction (1.7) one may write:

$$\text{Rate of reaction} = -\frac{\mathrm{d}[\mathrm{A}]}{\mathrm{d}t} = \frac{\mathrm{d}[\mathrm{B}]}{\mathrm{d}t} \tag{1.8}$$

In many instances a more careful definition of reaction rate is necessary, as is readily illustrated. For reaction (1.3), taking place in homogeneous aqueous solution, the rates of decrease of concentration of bromate and bromide ions are unequal and neither is equal to the rate of increase of the concentration of Br_2. The convention is that into each rate of change expressions we put, as a quotient, the corresponding stoichiometric coefficient in the balanced chemical equation, where the latter is written in the simplest manner that avoids fractions among these coefficients. Thus, for reaction (1.3), we have:

$$\text{Rate of reaction} = -\frac{\mathrm{d}[\mathrm{BrO_3^-}]}{\mathrm{d}t} = -\frac{1}{5}\frac{\mathrm{d}[\mathrm{Br^-}]}{\mathrm{d}t} = -\frac{1}{6}\frac{\mathrm{d}[\mathrm{H^+}]}{\mathrm{d}t} = \frac{1}{3}\frac{\mathrm{d}[\mathrm{Br_2}]}{\mathrm{d}t} \tag{1.9}$$

It is prudent to add that, in regard to reaction rate, we are speaking of the change of concentration *due to reaction.* Concentrations may, of course, be altered in other ways, for example by adding more of a substance or by diluting the reaction mixture with solvent. While the definition in equation (1.8) is ideal for a closed system, it needs to be used carefully if the change in concentration of a reactant contains contributions that do not arise from its reaction in the process under study.

For a reaction far from equilibrium, at constant temperature, the rate usually depends on the concentration of each reactant species, in the sense that the rate of reaction is proportional to the concentration raised to a certain power. Expressing it generally, we have the following definition. For the reaction represented by the stoichiometric equation,

$$\alpha A + \beta B + \gamma C = \lambda L + \mu M \qquad (1.10)$$

if the rate of reaction, $\frac{-1}{\alpha}\frac{d[A]}{dt}$, is proportional to $[A]^p[B]^q[C]^r$, then

p = order of reaction with respect to A,
q = order of reaction with respect to B,
r = order of reaction with respect to C,

and $n = p + q + r$ = the overall reaction order.

Two points implicit in this statement may merit special attention. Firstly, the rate of a chemical reaction is usually a function of the concentration of each *reactant,* but it is unusual for it to be influenced by the concentration of a reaction *product*. Secondly, the reaction orders, p, q and r, are not to be confused with the stoichiometric coefficients, α, β and γ, in the sense that the former group are neither required nor expected to be identical to the latter group.

To turn a proportionality into an equation merely requires the insertion of a constant of proportionality. By putting in such a constant we obtain the equation:

$$\text{Reaction rate} = \frac{1}{\alpha}\frac{d[A]}{dt} = k_n[A]^p[B]^q[C]^r \qquad (1.11)$$

This constant k_n is an nth order rate constant*, since n is the overall order of the reaction. The magnitude of k_n may be seen as a measure of the rapidity of the reaction. To consider the simplest situation, if the concentrations of A, B and C were all equal to unity then the reaction rate would be *numerically* equal to k_n: this illustrates the role of a rate constant as a criterion of the reaction rate under specified conditions.

The dimensions and units of the rate constant merit particular attention. Referring to equation (1.11), the left hand side has the dimensions of

* In 1887, Arrhenius argued that the constant of proportionality should be called the "specific reaction rate". Whatever the merits of his argument, his view did not prevail. Among practising kineticists and in the major journals of physical chemistry, the term "rate constant" has virtually universal acceptance, and so it is employed in this book.

Table 1.1 The dependence on the reaction order of the units of a rate constant and of the function of concentration that equates to kt.

Order of reaction	Units of the rate constant	Function of concentration equal to kt
0: zeroth order	mol dm^{-3} s^{-1}	x
1: first order	s^{-1}	$\ln\{a/(a-x)\}$
2: second order	dm^{3} mol^{-1} s^{-1}	$\dfrac{x}{a(a-x)}$
3: third order	dm^{6} mol^{-2} s^{-1}	$\dfrac{a^2-(a-x)^2}{2a^2(a-x)^2}$

concentration × (time)$^{-1}$. Since the right hand side must have the same dimensions and the product, $[A]^p[B]^q[C]^r$, has dimensions, (concentration)^n, it follows that the dimensions of the nth order rate constant k_n are (concentration)$^{(1-n)}$ × (time)$^{-1}$.

Although reaction order is defined as above, in a purely empirical manner, it is not unusual to find that it is a positive integer. If $n = 1$, the reaction is said to be *first* order; if $n = 2$, it is *second* order and if $n = 0$, it is *zeroth* order. The resulting units of the rate constant for these and other cases are shown in Table 1.1, where the unit of concentration employed is mol dm^{-3} which is that normally used for kinetic studies in solution.

1.3 Integrated rate equations

On the above basis, for any chemical reaction one can conceive of a rate equation analogous to equation (1.11). Since the left hand side of such an expression will necessarily involve the first derivative of concentration with respect to time, it follows that integration is essential to obtain an expression which inter-relates reactant concentration and time.

For simplicity, let us initially consider a first order reaction. Suppose the chemical reaction denoted by the equation

$$A = L \tag{1.10a}$$

is kinetically first order. Also, suppose the initial concentration of A at $t = 0$ is equal to a: it is convenient to introduce the variable x to denote the change in concentration as reaction proceeds. So at time t after the reaction has commenced, the concentration of A has fallen to $(a-x)$ and that of L has risen to x. We then have, corresponding to equation (1.11) the relation:

$$-\frac{\mathrm{d}(a-x)}{\mathrm{d}t} = \frac{\mathrm{d}x}{\mathrm{d}t} = k_1(a-x) \tag{1.12}$$

Rearranging the equation to separate the variables, x and t, we obtain:

$$\frac{\mathrm{d}x}{(a-x)} = k_1 \mathrm{d}t \tag{1.13}$$

Integrating each side of this equation, we obtain:

$$\ln(a-x) = k_1 t + \text{const} \tag{1.14}$$

The constant of integration is easily identified. When $t = 0$, $x = 0$, and so the constant is equal to $-\ln a$. This leads to:

$$-\ln a - \ln(a-x) = \ln\left\{\frac{a}{a-x}\right\} = k_1 t \tag{1.15}$$

Equation (1.15) shows that $\ln(a-x)$ will be a linear function of t, that is, the logarithm of the reactant concentration decreases linearly with time. Also, by taking antilogarithms of each side we obtain the equation,

$$(a-x) = a \exp(-k_1 t) \tag{1.16}$$

which indicates that $(a-x)$ decreases exponentially with time. These relations are illustrated diagrammatically in Figure 1.1.

To consider the integration of the rate equation for a second order reaction, let us assume that the reaction is first order with respect to each of two reactants, A and B. Further, let us assume (for simplicity) that the reaction involves equal proportions of each reactant, i.e. that A and B react in the ratio of one molecule of A to one molecule of B.

If we denote reactant concentrations as follows:

	A	B
$t = 0$	a	b
$t = t$	$(a-x)$	$(b-x)$

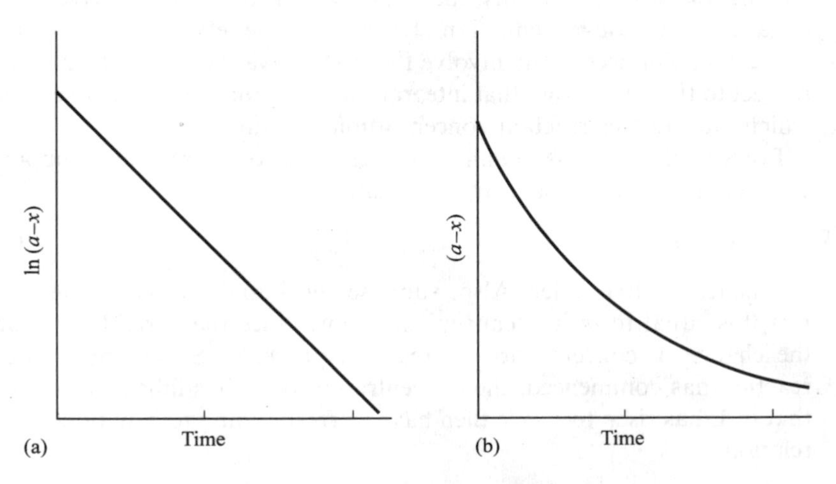

Fig. 1.1 A sketch of the significance of equations (1.16) and (1.17): (a) a plot of $\ln(a-x)$ against t and (b), of $(a-x)$ against t.

then the rate equation can be written,

$$\frac{\mathrm{d}x}{\mathrm{d}t} = k_2(a - x)(b - x) \tag{1.17}$$

Separating the variables as before leads to the relation:

$$\frac{\mathrm{d}x}{(a - x)(b - x)} = k_2\mathrm{d}t \tag{1.18}$$

If we consider first the special case where the initial concentrations of the reactants A and B are equal, i.e. where $a = b$, then we have:

$$\frac{\mathrm{d}x}{(a - x)^2} = k_2\mathrm{d}t \tag{1.19}$$

This is easily integrated and using the initial condition as before to identify the constant of integration, it leads to the following expression for the integrated rate equation:

$$\frac{1}{(a - x)} - \frac{1}{a} = k_2 t \tag{1.20}$$

Thus in this situation, the reciprocal of $(a - x)$, the reactant concentration, increases linearly with t.

Reverting to equation (1.18), let us now address the general case where the concentrations of A and B are initially unequal, that is $a \neq b$. To integrate this equation it is desirable to find constants D and E such that

$$\frac{1}{(a - x)(b - x)}$$

may be represented as

$$\frac{D}{(a - x)} + \frac{E}{(b - x)}$$

Simple algebra identifies the appropriate constants as $D = (b - a)^{-1}$ and $E = (a - b)^{-1}$. So equation (1.18) becomes

$$\frac{1}{(b - a)}\left\{\frac{\mathrm{d}x}{(a - x)} - \frac{\mathrm{d}x}{(b - x)}\right\} = k_2\mathrm{d}t \tag{1.21}$$

Integrating this equation in the same manner as equation (1.13) and applying the familiar initial condition to the constant of integration, leads to the relation:

$$\frac{1}{(b - a)} \ln\left\{\frac{a(b - x)}{b(a - x)}\right\} = k_2 t \tag{1.22}$$

An interesting case arises when b is not simply greater than a, but is very much greater, perhaps by a factor of 20 or more. Since x can never exceed a, then $(b - a)$ and $(b - x)$ are scarcely distinguishable from b and so the equation reduces to:

$$\ln\left\{\frac{a}{a-x}\right\} = bk_2t \tag{1.23}$$

This has a form similar to equation (1.15) and indicates that under these conditions, where B is present in large excess, the concentration of A should decrease exponentially; in other words, this second order reaction now exhibits first order characteristics. We say that the reaction has been rendered *pseudo-first* order since it follows an expression resembling a first order rate equation. Furthermore, the analogue in equation (1.23) of k_1 in equation (1.15) is called the *pseudo-first* order rate constant, viz, k_2b. This means that when a reaction which is kinetically first order with respect to each of two reactants is taking place with one reactant present in large excess, the concentration of the other will decrease exponentially but at a rate which is proportional to the concentration of the reactant in excess.

Since the order of a reaction might conceivably have any value, integral or non-integral, it is desirable to consider the general case of a reaction of nth order, i.e. the rate equation is

$$-\frac{1}{\alpha}\frac{d[A]}{dt} = k_n[A]^p[B]^q[C]^r \tag{1.24}$$

where $(p+q+r)=n$. We wish to look at the case where the various reactants are present in stoichiometric proportion, that is the ratio of the initial concentrations, $a:b:c$ is the same as that of the stoichiometric coefficients, $\alpha:\beta:\gamma$. If these coefficients are all unity then we have equation (1.25):

$$-\frac{d[A]}{dt} = k_n[A]^n \quad \text{or} \quad \frac{dx}{dt} = k_n(a-x)^n \tag{1.25}$$

(More generally, we will also have on the right hand side a constant involving α, β, γ, q and r.) Where $n \neq 1$, this leads on integration to the equation:

$$\frac{1}{(n-1)(a-x)^{n-1}} - \frac{1}{(n-1)a^{n-1}} = k_nt \tag{1.26}$$

A very well-known example of this general relationship is the integrated second order rate equation (1.20), which we derived above. Another example is the zero order equation:

$$a-(a-x) = k_0t \tag{1.27}$$

These equations exemplify the general rule implicit in equation (1.26) that for a reaction of nth order, with all reactants present in stoichiometric ratio, the reactant concentration to the power $(1-n)$ will vary linearly with time.

As an illustration of the effect that reaction order has on the way reactant concentration is depleted, Figure 1.2 demonstrates the differing patterns for a zero order, a first order and a second order reaction, where in all cases the initial rate is identical. When the zero order reaction is complete, the first order reaction has still to go $1/e$ of the way to completion and the second order reaction has reached half-way. The difference in behaviour between

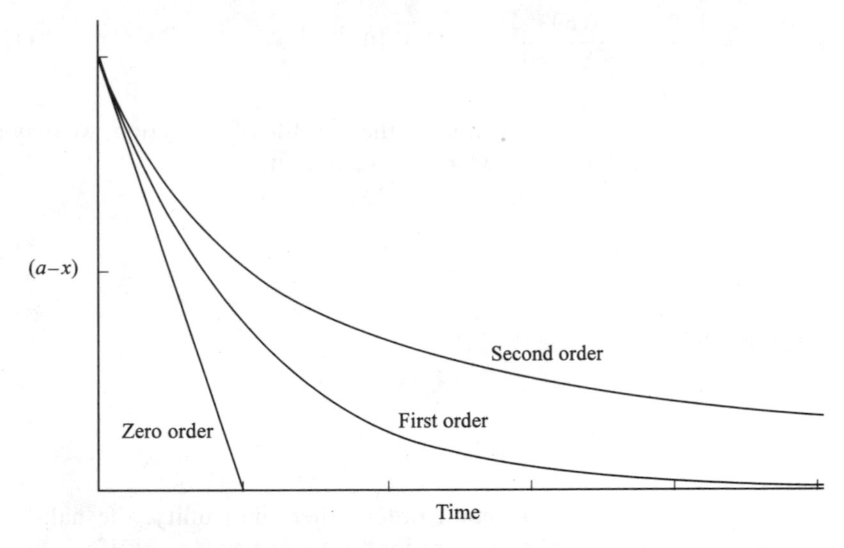

Fig. 1.2 Representation of the progress of reactions of differing order, all with the same initial concentration and initial rate.

the first order and the second order reactions becomes increasingly striking after that point.

1.4 Reaction half-life and mean lifetime

To indicate the rapidity of rate processes, chemists often use the concept of half-life, defined as the time required for a reaction to proceed half-way to completion and usually denoted by $t_{1/2}$. So if the initial concentration is a, at $t_{1/2}$ it has fallen to $a/2$.

Starting with the first order case, by substituting as above into equation (1.15) we obtain

$$\ln 2 = k_1 t_{1/2} \tag{1.28}$$

or

$$t_{1/2} = \frac{\ln 2}{k_1} \tag{1.29}$$

Thus for a first order process, the half-life depends only on the first order rate constant and not in any way on the initial concentration. This fact is explicitly used in regard to radioactive decay, which is a prime example of a first order process. Conventionally, the rate at which nuclei of an unstable nuclide disintegrate is tabulated, not as the value of the relevant first order rate constant but as that of the half-life. For example, for $^{221}_{87}$Fr, $t_{1/2} = 4.8$ min. From this we can calculate the rate constant:

$$k_1 = \frac{\ln 2}{t_{1/2}} = \frac{0.693}{4.8 \times 60} = 2.41 \times 10^{-3}\ \mathrm{s}^{-1} \tag{1.30}$$

To derive a general relationship for the half-life of a reaction, we may use equation (1.26). Putting $(a - x) = a/2$ we obtain,

$$\frac{2^{n-1} - 1}{(n-1)a^{n-1}} = k_n t_{1/2} \tag{1.31}$$

which leads to:

$$t_{1/2} = \frac{2^{n-1} - 1}{k_n (n-1) a^{n-1}} \tag{1.32}$$

This shows that for any reaction order other than unity, the half-life is proportional to the initial concentration to the power $(1 - n)$. If n is greater than 1, then $t_{1/2}$ becomes less the higher the initial concentration; for n less than 1 the reverse is true.

A related but different quantity is the mean lifetime of a reactant molecule, that is, the mean of the lifetimes of all molecules undergoing reaction. If the concentration of these at any time t is $(a - x)$, then the mean lifetime, τ, is $\int_0^\infty (a - x)\mathrm{d}t$ divided by the initial concentration, a.

For a zero order reaction with a rate constant k_0, the concentration at time t will be $(a - x) = a - k_0 t$. This means the concentration $a/2$ will be attained in a time $a/2k_0$. For the mean lifetime we have:

$$\tau = \frac{\int_0^{a/k_0} (a - k_0 t)\mathrm{d}t}{a} = \frac{a}{2k_0} \tag{1.33}$$

Thus, as illustrated in Figure 1.3, for a zero order reaction the half-life and the mean lifetime are the same.

For a reactant whose decay follows first order kinetics, we have,

$$\tau = \frac{\int_0^\infty a\,\mathrm{e}^{-k_1 t}}{a} = \frac{1}{k_1} \tag{1.34}$$

This is, in fact, the time required for the concentration to decrease to 1/e of its initial value and is greater than the half-life which, as shown above is $0.693/k_1$ (Figure 1.4).

For a second order reaction, the rate of decay slows down with time to an even greater extent than for a first order reaction. The mean lifetime is given by:

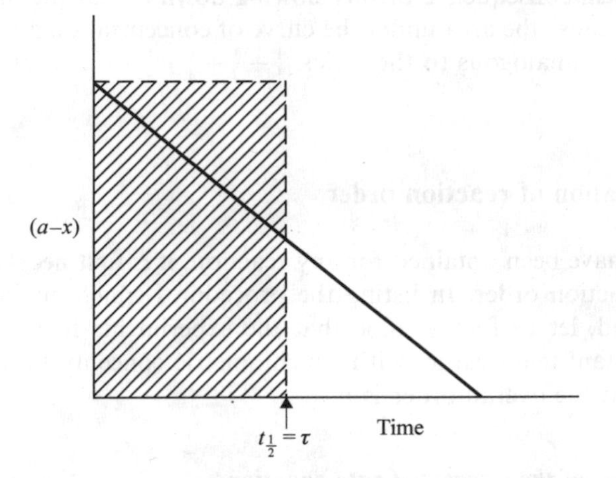

Fig. 1.3 Diagrammatic demonstration that, for a zero order reaction, the half-life and the mean lifetime coincide. It can readily be seen that the area under the solid curve is equal to the shaded area within the dotted rectangle.

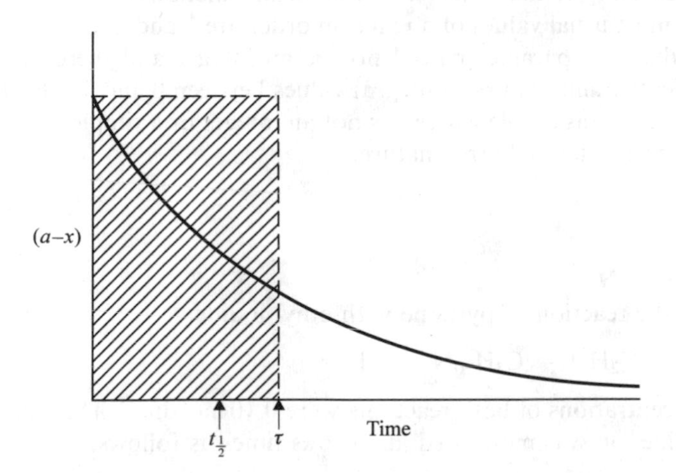

Fig. 1.4 Diagrammatic comparison of the half-life and the mean lifetime for a first order reaction. The area under the solid curve, continued on to infinity, is equal to the shaded area within the dotted rectangle.

$$\tau = \frac{\displaystyle\int_0^\infty \mathrm{d}t \Big/ \left(\frac{1}{a} + k_2 t\right)}{a}$$

$$= \frac{\displaystyle\frac{1}{k_2}\left[\ln\left(k_2 t + \frac{1}{a}\right)\right]_0^\infty}{a} = \infty \qquad (1.35)$$

Thus the mathematical consequence of this slowing down is that the mean lifetime is infinite, because the area under the curve of concentration against time has no limit. It is analogous to the series, $\frac{1}{1}+\frac{1}{2}+\frac{1}{3}+\frac{1}{4}+\frac{1}{5}\ldots$, which does not converge.

1.5 The determination of reaction order

Where kinetic data have been obtained for any reaction, it is first necessary to determine the reaction order. In listing the procedures which might be employed to this end, let us first assume that the experiments have been performed at a constant temperature with stoichiometric concentrations of all reactants and that the overall order is n.

1.5.1 Method (i): use of the integrated rate equation

As shown in equations (1.26), (1.15) and (1.20), if $(a - x)$ is a linear function of t (i.e. it decreases at a constant rate) then $n = 0$; if $\ln(a - x)$ is a linear function of t then $n = 1$; and if $(a - x)^{-1}$ is a linear function of t then $n = 2$. Although the most usual values of a reaction order are 1 and 2, it should be remembered that this parameter need not be an integer and there are, of course, an infinite number of non-integral values between 0 and 3. This fact renders the method unsuitable where n is not an integer or a simple fraction, in view of its basic trial-and-error nature.

Example 1.1

In a study of the reaction of pyridine with ethyl iodide,

$$C_5H_5N + C_2H_5I = C_7H_{10}N^+ + I^-$$

the initial concentrations of both reactants were 0.10 mol dm^{-3}. The concentration of iodide ion was monitored at various times as follows:

t/s	230	465	720	1040	1440	1920	2370
$[I^-]$/mol dm^{-3}	0.015	0.026	0.035	0.044	0.052	0.059	0.064

Determine the reaction order and the rate constant.

Answer. It is important to realise that the tabulated concentrations refer to I^- ion, which is a *product* of the reaction. These concentrations must first be used to find those of pyridine and ethyl iodide, since the equations referred to above all relate to *reactant* concentration. Thus we have, using the information given and the stoichiometry of the reaction.

t/s	0	230	465	720	1040	1440	1920	2370
$[C_5H_5N]$/mol dm^{-3}	0.100	0.085	0.074	0.065	0.056	0.048	0.041	0.036

To find the reaction order, let us test if $n = 1$ and if $n = 2$. This involves finding $\ln[C_5H_5N]$ and $1/[C_5H_5N]$.

t/s	0	235	465	720	1040	1440	1920	2370
$\ln\{[C_5H_5N]/\text{mol dm}^{-3}\}$	−2.30	−2.47	−2.60	−2.73	−2.88	−3.04	−3.19	−3.32
$[C_5H_5N]^{-1}/\text{dm}^3\,\text{mol}^{-1}$	10.0	11.8	13.5	15.4	17.9	20.8	24.4	27.-8

When these are plotted against t, the results are as shown in Figure 1.5. Clearly, the plot of $\ln[C_5H_5N]$ is curved so the reaction does not follow first order kinetics. Since the plot of $1/[C_5H_5N]$ against t is linear, we can say that kinetically the reaction is second order, with a rate constant, as given by the slope of the graph, of $7.52 \times 10^{-3}\ \text{dm}^3\ \text{mol}^{-1}\ \text{s}^{-1}$.

Traditionally, the plotting of the relevant graph has been the means used to test that a set of data fits a certain equation. The phenomenal increase in recent years in the power and availability of electronic calculators has meant that a non-graphical approach may be equally reliable and much faster. For example, the second order integrated rate equation (1.20) above, may be rearranged to give equation (1.36).

$$k_2 = \frac{1}{t}\frac{x}{a(a-x)} \tag{1.36}$$

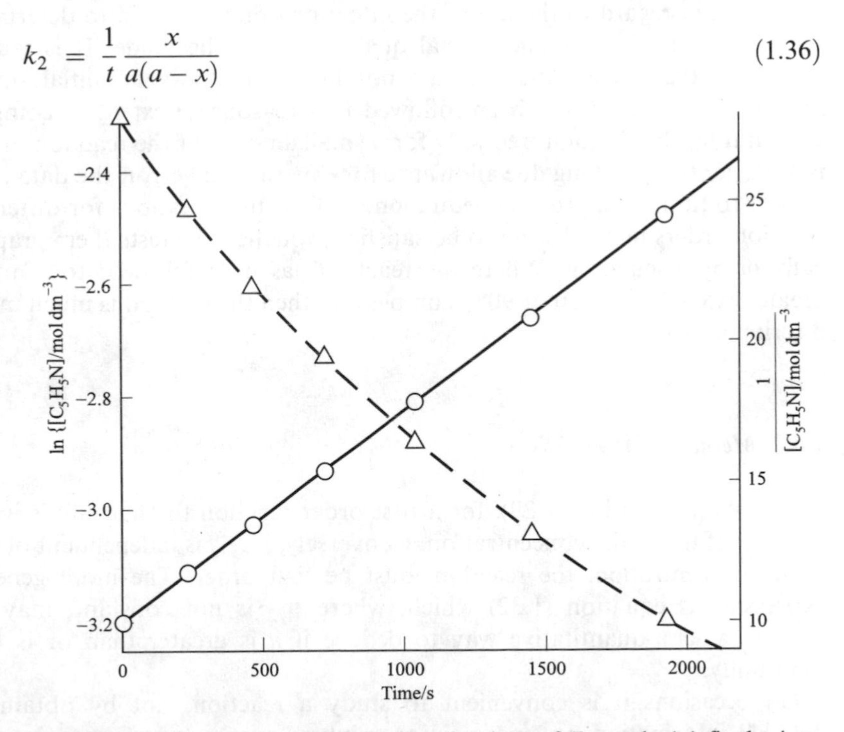

Fig. 1.5 Graphical tests as to whether the data of Example 1.1 fit the integrated rate equation of a first order reaction (Δ. l.h. scale) or a second order reaction (○, r.h. scale).

Applying this relation to the data of Example 1.1, it may readily be shown that they yield the values 7.51×10^{-3}, 7.56×10^{-3}, 7.49×10^{-3}, 7.55×10^{-3}, 7.52×10^{-3}, 7.50×10^{-3} and $7.50 \times 10^{-3}\ dm^3\ mol^{-1}\ s^{-1}$. Since these are (to two significant figures) constant values, it follows that the graph of $1/[C_5H_5N]$ against t is linear, that the reaction is second order and that the constant value obtained above is indeed the rate constant of this reaction.

Alternatively we might try, in a similar way, to fit the data to the integrated first order equation, i.e. to evaluate the right hand side of the expression,

$$k_1 = \frac{1}{t}\ln\left\{\frac{a}{a-x}\right\} \tag{1.37}$$

obtained by rearranging equation (1.16). This yields the values 6.91×10^{-4}, 6.48×10^{-4}, 5.98×10^{-4}, 5.58×10^{-4}, 4.64×10^{-4} and $4.31 \times 10^{-4}\ s^{-1}$, and since these are not constant but are steadily decreasing for increasing value of t, it follows that the plot of $\ln[C_5H_5N]$ against t is not a linear one, that the reaction is not first order and that none of the numbers obtained in evaluating the right hand side of equation (1.37) has any significance in the context of this reaction.

Finally, in regard to the use of the integrated rate equation to determine reaction order there is one crucial qualification to be made. It is vitally important that the kinetic data are not limited to the very initial stages and that the reaction has been followed to a reasonable extent of completion. If it has been monitored only for a small amount of the reaction, there is a danger that, making due allowance for experimental error, the data may appear to fit an inappropriate equation. In fact, the equations for different reaction orders may all seem to be satisfied, whether one tests them graphically or by calculation. Where the reaction has been followed to a much greater extent, e.g. to 70 or 90% completion, then the method is much more discriminatory.

1.5.2 Method (ii): half-lives

As shown in equation (1.29), for a first order reaction the half-life is independent of the initial concentration. Conversely, if $t_{1/2}$ is independent of the initial concentration, the reaction must be first order. The more general expression is equation (1.32) which, where $t_{1/2}$ is not constant, may be used in a semi-quantitative way to deduce if n is greater than or is less than unity.

On occasions it is convenient to study a reaction, not by obtaining detailed information on one reaction mixture over a long period but by finding for a series of reaction mixtures the time required for the reaction to go a certain fraction of the way to completion. This might arise if the reaction may most readily be followed by a "clock" type method (see Section 2.2).

By taking natural logarithms of both sides of equation (1.32) we obtain equation (1.38)

$$\ln t_{1/2} = \ln \frac{(2^{n-1} - 1)}{k_n(n-1)} + (1-n)\ln a \tag{1.38}$$

It may be noted that the first term on the r.h.s. of this equation is constant. By plotting ln $t_{1/2}$ against ln a, we can find n directly.

Example 1.2

For the thermal decomposition of N_2O at 1030 K, the half-life, $t_{1/2}$, was found to vary as follows with the initial pressure, P_0:

P_0/torr	86.5	164	290	360
$t_{1/2}$/s	634	393	255	212

Deduce the order of the thermal decomposition of N_2O.

Answer. The initial pressure, P_0 is of course a measure of the initial concentration of N_2O. And since we are going to take logarithms of each value, there is no need to convert from torr into any other unit of concentration. Thus we have:

$\ln(P_0/\text{torr})$	4.46	5.10	5.67	5.89
$\ln(t_{1/2}/\text{s})$	6.45	5.97	5.54	5.36

The inter-relation of these is shown in Figure 1.6, which yields a gradient of −0.76. Equating this to $(1 - n)$, we have $n = 1.76$.

1.5.3 Method (iii): initial rates

In our discussions to date, it has been assumed that the only factor contributing to a change in the reaction rate as the reaction proceeds is the changing concentrations of reactants. On occasions, other factors may also be operative, such as a growing importance of the reverse reaction or, in heterogeneous catalysis, sintering or poisoning of the catalyst. These factors will be minimised in the initial stages.

If we consider equation (1.25) as referring to the initial stage of a reaction, by taking logarithms of both sides we obtain:

$$\ln(\text{Initial rate}) = \ln k_n + n\ln[\text{A}]_0 \tag{1.39}$$

This means that if the initial rate is measured as a function of reaction concentration, the overall order may readily be found by taking the gradient of the relevant plot.

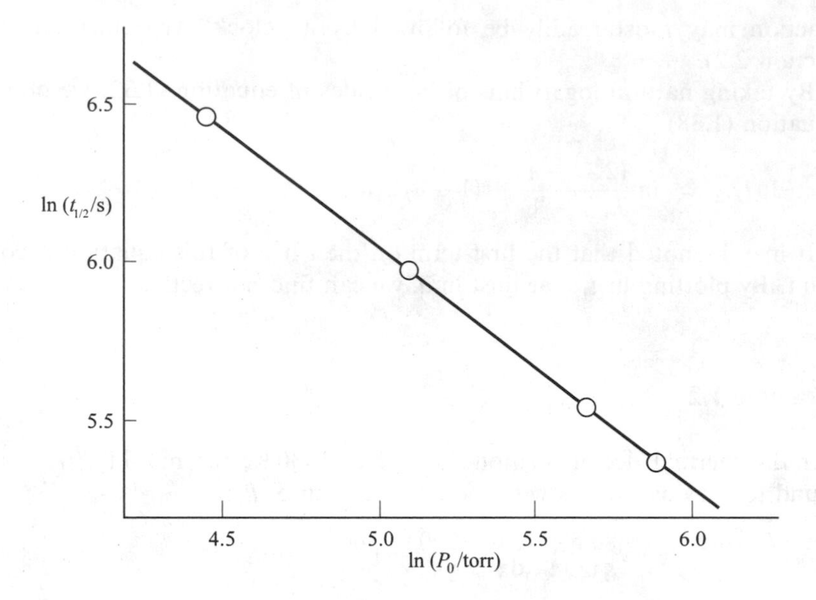

Fig. 1.6 Plot of $\ln t_{1/2}$ against ln P_0 using the data of Example 1.2.

In the foregoing discussion, it has been assumed that the concentrations of all reactants are maintained in stoichiometric proportion. The overall order, n, may then be found by method (i), (ii) or (iii). The question then arises, what are the individual orders of reaction, p, q, and r, with respect to each reactant?

The general approach has been outlined in discussing a reaction which is kinetically first order with respect to each of two reactants. Let us start from equation (1.24). If B is present in large (e.g. 20-fold) excess, then its concentration scarcely changes during the course of the reaction. If A and C are present in stoichiometric ratio then we have,

$$-\frac{\mathrm{d[A]}}{\mathrm{d}t} = k_n[\mathrm{B}]_0^q[\mathrm{A}]^{p+r} \tag{1.40}$$

which means that the reaction will behave as if it is of order $(p + r)$.

From kinetic data obtained using this large excess of B, one might then determine $(p + r)$, by methods (i) or (ii). Similarly, if both B and C are present in large excess, then the reaction will be *pseudo-p^{th}* order. This device, which enables the individual orders with respect to each reactant to be determined is called the *isolation method.*

If only the initial rate is being used to find reaction order, then the situation is slightly different. Taking logarithms of each side of equation (1.24), we obtain

$$\ln(\text{Initial rate}) = \ln k_n + p\ln[\mathrm{A}]_0 + q\ln[\mathrm{B}]_0 + r\ln[\mathrm{C}]_0 \tag{1.41}$$

Thus to determine p, it is merely required to find how the initial rate varies with the concentration of A, keeping constant those of B and C; it is not also necessary that B and C be present in large excess.

Example 1.3

The reaction of nitric oxide with hydrogen,

$$2NO + 2H_2 = N_2 + 2H_2O \tag{1.42}$$

has been studied by following the change in total pressure. At a fixed hydrogen pressure the following rate data were obtained:

NO pressure/torr	359	300	240	183
Initial rate/torr s^{-1}	1.50	1.03	0.64	0.37

Answer. Assuming orders p and q with respect to NO and H_2 respectively, we have

$$\text{Initial rate} = k_n(P_{NO})^p(P_{H_2})^q \tag{1.43}$$

$$\therefore \ln(\text{Initial rate}) = \ln k_n + p \ln P_{NO} + q \ln P_{H_2} \tag{1.44}$$

We wish to plot the logarithm of the initial rate against the logarithm of the initial pressure of NO.

$\ln(P_{NO}/\text{torr})$	5.88	5.70	5.48	5.22
ln (Initial rate)	0.40	0.03	−0.45	−0.99

From the plot in Figure 1.7, it may be seen that the gradient is 2.08, which is thus the experimental value for p, the order with respect to NO.

1.6 Effect of temperature on reaction rates

In general, the rate of a chemical reaction is accelerated by an increase in temperature. The effect is normally represented in terms of the variation with temperature of the rate constant, k, which frequently follows the Arrhenius equation:

$$k = A\,e^{-E_a/RT} \tag{1.45}$$

An alternative form is obtained by taking the logarithm of each side.

$$\ln k = \ln A - E_a/RT \tag{1.46}$$

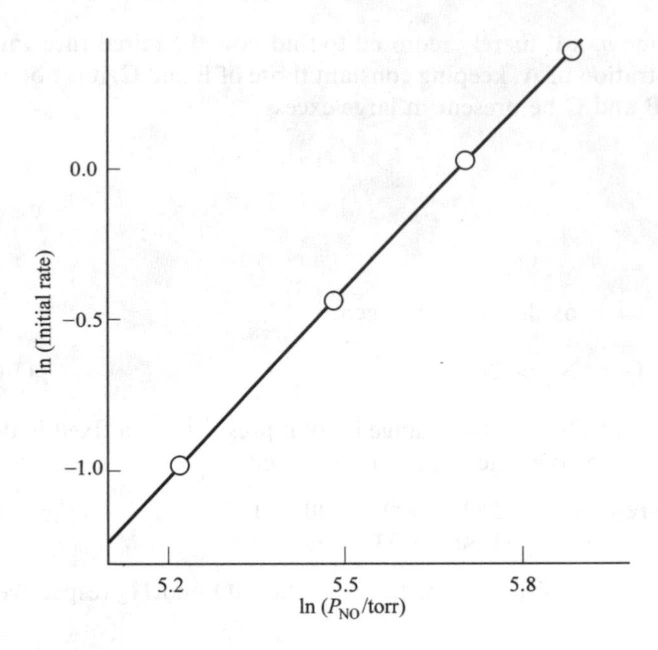

Fig. 1.7 Plot of ln(Initial rate) against ln P_{NO} using the data of Example 1.3.

This implies that the logarithm of the rate constant is a linear function of the reciprocal temperature. Such a graph is often referred to as an Arrhenius plot.

Of the two parameters, both assumed independent of temperature, introduced in equation (1.45), A is called the pre-exponential factor or the A factor and always has the same units as the corresponding rate constant. The other, E_a, known as the activation energy or more properly as the Arrhenius activation energy, has the same dimensions as RT and typical values may lie in the range 50–200 kJ mol^{-1}.

Example 1.4

For a decomposition reaction, the first order rate constant is found to vary with temperature as follows:

T/°C	5	15	25	35
k/s^{-1}	1.5×10^{-6}	8.0×10^{-6}	4.1×10^{-5}	2.0×10^{-4}

Construct an Arrhenius plot and thus determine the Arrhenius parameters E_a and A for this reaction.

Answer. This is an invitation to fit the data to equation (1.46). So we want to plot ln k against T^{-1}.

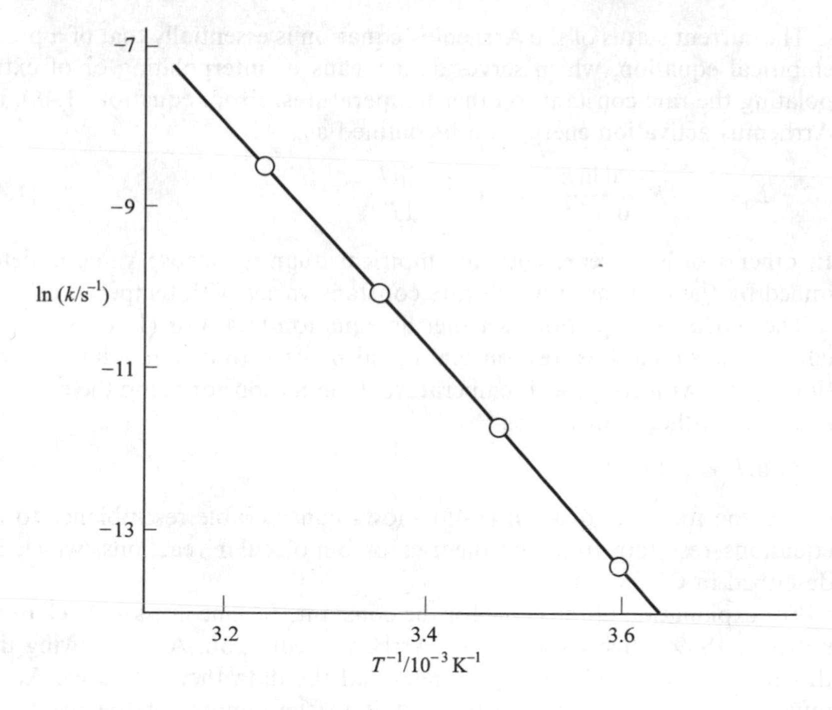

Fig. 1.8 Plot of $\ln k$ against T^{-1} using the data of Example 1.4.

T/K	278.1	288.1	298.1	308.1
$\ln(k/\text{s}^{-1})$	−13.41	−11.74	−10.10	−8.52
$T^{-1}/10^{-3}\,\text{K}^{-1}$	3.595	3.470	3.354	3.245

This plot is shown in Figure 1.8, which shows a good linear relationship. Equating the slope to $-E_a/R$, we have

$$\begin{aligned} E_a &= -R \times \text{slope} \\ &= -8.314 \times \frac{(-13.48 - (-9.31))}{(3.60 - 3.30) \times 10^{-3}} \\ &= 115.6 \times 10^3\,\text{J mol}^{-1} \quad \text{or} \quad 115.6\,\text{kJ mol}^{-1} \end{aligned}$$

The easiest way to evaluate A is by substitution into equation (1.46), now that E_a has been determined

$$\ln(A/\text{s}^{-1}) = \ln(k/\text{s}^{-1}) + \frac{E_a}{RT}$$

Using the data at 15°C, and keeping E_a in the same units as RT, we have

$$\begin{aligned} \ln(A/\text{s}^{-1}) &= -11.74 + \frac{115.6 \times 10^3}{8.314 \times 288.1} = 36.50 \\ \therefore A &= 7.1 \times 10^{15}\,\text{s}^{-1} \end{aligned}$$

The current status of the Arrhenius equation is essentially that of a purely empirical equation, which serves as a means of interpolating or of extrapolating the rate constant to other temperatures. From equation (1.46), the Arrhenius activation energy can be defined as:

$$E_a = -R \cdot \frac{d \ln k}{d(1/T)} = RT^2 \frac{d \ln k}{dT} \tag{1.47}$$

In other words, it represents an empirical quantity whose value is determined by the manner in which rate constant varies with temperature.

The Arrhenius equation, whether as equation (1.45) or (1.46), is not, of course, the simplest expression that could be used to denote a linear variation of $\ln k$ with reciprocal temperature. One reason for using the equation as above, rather than writing:

$$\ln k = A + B/T \tag{1.48}$$

is that the form of equation (1.45) shows considerable resemblance to the equations resulting from the theories of bimolecular reactions, which are described in Chapter 4.

For explanation of the name for the constant, E_a, one must go back to the paper in 1889 which launched the Arrhenius equation. After showing that this relation could adequately correlate all the data then available, Arrhenius then presented a justification of it in the context of the hydrolysis reaction:

$$\text{sucrose} + H_2O = \text{glucose} + \text{fructose} \tag{1.49}$$

He postulated that this involved an isomerism between "active" and "inactive" forms of sucrose, with only the former able to undergo hydrolysis.

$$\text{"inactive" sucrose} \rightleftharpoons \text{"active" sucrose} \tag{1.50}$$

$$\text{"active" sucrose} + H_2O \rightarrow \text{glucose} + \text{fructose} \tag{1.51}$$

Also, he implicitly assumed (i) that the isomerisation process (1.50) is sufficiently rapid that the reaction of the active form does not deplete the concentration from the equilibrium value and (ii) that the rate of reaction of the active form is independent of temperature.

On this basis, the rate constant of reaction (1.49) would be proportional to K_c, the equilibrium constant for process (1.50), and thus to $\exp(-\Delta H^{\ominus}/RT)$, where $\Delta H^{\ominus}$ denotes the standard enthalpy change for this reaction. Not unnaturally, given the labels attached to the isomers involved in this process, he called this quantity "the activation energy", and the name endures even though the steps on which it is based have long been discarded.

While equation (1.45) is one of the most widely used expressions of reaction kinetics, in many instances the rate constant is found to vary with temperature in a way that shows appreciable deviation from this behaviour.

Consequently, other equations have been proposed, with three or more parameters. One of these is the relation:

$$\ln k = A + \frac{B}{T} + C \ln T \tag{1.52}$$

which has been widely used, especially with reactions in solution. Another more recent equation is

$$\ln k = A + BT^{C} \tag{1.53}$$

where better fit than is achieved by the Arrhenius equation may be found with C put equal to non-integral values, differing somewhat from -1.

In some cases, the rate constant for a reaction may show quite considerable deviation from the Arrhenius equation, in a way that merits a special explanation. Examples of such behaviour will be referred to in other chapters where the reason for the errant behaviour will be examined.

Suggested reading

Van't Hoff, J. H., 1884. *Études de dynamique chimique*. Amsterdam: F. Muller and Co.

Harcourt, A. V., and W. Esson, 1886. On the laws of connexion between the conditions of a chemical change and its amount, Part 1. *Philosophical Transactions of the Royal Society,* **156**, 193–221 and especially the Appendix by Esson, W.

Arrhenius, S. A., 1889. Über die Reaktionsgeschwindigkeit bei der Inversion von Rohrzucker durch Säuren. *Zeitschrift für physikalische Chemie,* **4**, 226–48.

King, M. C., 1981. Experiments with time: progress and problems in the development of chemical kinetics. *Ambix*, **28** (2), 70–82; **29** (1), 49–61.

Logan, S. R., 1984. Introductory reaction kinetics – an unacknowledged difficulty. *Education in Chemistry*, **21**, 20–22.

Laidler, K. J., 1985. Chemical kinetics and the origins of physical chemistry. *Archive for History of Exact Sciences,* **32**, 43–75.

Laidler, K. J., 1984. The development of the Arrhenius equation. *Journal of Chemical Education,* **61**, 494–98.

Logan, S. R., 1982. The origin and status of the Arrhenius equation. *Journal of Chemical Education*, **59**, 279–81.

Problems

1.1 The pyrolysis of dimethyl mercury was found to be a first order reaction. A sample of the substrate was held at a constant temperature for a known period of time, after which the extent of decomposition was determined. From the following data [Cattenach, J. and L. H. Long, *Trans Faraday Soc.,* 1960, **56**, 1286–95], deduce the decomposition rate constant at each temperature and thus evaluate the Arrhenius parameters, E_a and A.

T/°C	time/min	% decomposition
331.7	120	32.4
319.8	330	35.3
305.2	840	33.5

1.2 The concentration of benzyl radicals, $R^{\bullet}$, was found to decrease as follows after the pulse of electrons which caused this radical to be formed. Determine the order of the decay [McCarthy R. L., and A. MacLachlan, *Trans Faraday Soc.*, 1960, **56**, 1187–200] and the relevant rate constant.

$t/\mu s$	31	43	60	80	101	121
$[R^{\bullet}]/10^{-5}\,mol\,dm^{-3}$	7.14	6.53	5.38	4.55	3.82	3.45

$t/\mu s$	150	220	280	350	400
$[R^{\bullet}]/10^{-5}\,mol\,dm^{-3}$	2.80	1.97	1.55	1.26	1.15

1.3 In acidic aqueous solution, Ce^{IV} reacts with H_2O_2 to yield Ce^{III} and O_2. In an experiment in which the initial concentrations were $[Ce^{IV}] = 9.25 \times 10^{-5}\,mol\,dm^{-3}$ and $[H_2O_2] = 7.5 \times 10^{-6}\,mol\,dm^{-3}$, the concentration of hydrogen peroxide was found to fall with time as follows:

time/ms	2	4	6	8	10
$[H_2O_2]/10^{-6}\,mol\,dm^{-3}$	6.23	4.84	3.76	3.20	2.60

time/ms	12	14	16	18	20
$[H_2O_2]/10^{-6}\,mol\,dm^{-3}$	2.16	1.85	1.49	1.27	1.01

Show that the reaction is following pseudo-first order kinetics and evaluate the rate constant [Czapski, G., B. H. J. Bielski and N. Sutin, *J. Phys. Chem.*, 1963, **67**, 201–3]

1.4 In a study of the pyrolysis of 1-butene [Bryce, W. A., and P. Kebarle, *Trans Faraday Soc.*, 1958, **54**, 1660–77], methane was produced as a major product with the following values of the first order rate constant

T/°C	493	509	514	522	541	546	555
$k/10^{-5}\,s^{-1}$	8.4	24.1	24.2	38.1	90.2	140	172

Evaluate the Arrhenius parameters, E_a and A, of this reaction.

1.5 The rates of decarboxylation of the anions of substituted carboxylic acids were found to be proportional to the concentration of the anion and to that of H^+ ion [Willi, A. V., *Trans Faraday Soc.*, 1959, **55**, 433–41]. For 4-amino-salicylate, the rate constant varied with temperature as follows:

T/°C	20.07	36.87	50.03
$k/10^{-2}\,dm^3\,mol^{-1}\,s^{-1}$	1.80	13.1	43.0

Evaluate the Arrhenius parameters, E_a and A, of this reaction and deduce the value of the rate constant at 70°C.

1.6 The radionuclide ^{32}P has a half-life of 14.3 days. If one were to acquire a sample of this β^- emitter, how long would it take until the amount remaining is 1.0% of the initial sample?

1.7 In the reaction of HBr with NO_2 at 127°C

$$2HBr + NO_2 = H_2O + NO + Br_2$$

the following table shows the effect on the initial rate of varying the partial pressure of HBr while that of NO_2 was kept constant at 10 torr:

HBr pressure/torr	4.7	11.0	15.3	20.3	26.1	34.1
Initial rate/torr min^{-1}	0.98	2.35	3.16	3.96	5.56	6.77

The second table shows the effect of varying the partial pressure of NO_2 while that of HBr was kept constant at 10 torr:

NO_2 pressure/torr	5.0	10.2	15.4	20.3	25.7
Initial rate/torr min^{-1}	1.27	2.37	3.64	4.94	6.19

Evaluate the reaction order with respect to each reactant and the initial rate of reaction at 127°C when the partial pressures of both HBr and NO_2 are equal to 10 torr [Addecott, K. S. B., and J. H. Thomas, *Trans Faraday Soc.*, 1961 **57,** 664–69].

2 The experimental study of reaction kinetics

This chapter considers in a critical manner the factors that are important in designing experiments to obtain data on the kinetics of chemical reactions. It details a number of techniques and discusses their special merits.

Regarding the interpretation of experimental data, attention is given to techniques designed to help in deriving an accurate value of the relevant rate constant. In this connection, the increasing role of computers is carefully considered.

2.1 Minimum requirements for the study of chemical kinetics

In order for the study of the rate of a chemical reaction to be meaningful, it is in general necessary that there should be strict temperature control of the reacting mixture. Clearly this is on account of the appreciable temperature dependence of reaction rates. An additional reason is that many techniques used to monitor the progress of a reaction involve measuring a parameter whose value is intrinsically temperature-dependent.

Frequently, in the study of the kinetics of a reaction in solution, the reaction is commenced by mixing two or more solutions which together contain all the component species for the reaction. In such an instance it is important that the mixing is thorough, since achieving complete homogeneity by diffusion of the reactants is a notoriously slow process. The root mean square distance of diffusion, l, of a molecule in any direction is given by the equation,

$$l = \sqrt{6Dt} \tag{2.1}$$

where D is the diffusion coefficient and t is the interval of time. Let us take $D = 2 \times 10^{-9}\ \mathrm{m^2\ s^{-1}}$, which is a characteristic value for a solute in aqueous solution. If t is taken as 10 minutes = 600 s, then l is found to be $2.68 \times 10^{-3}\ \mathrm{m} = 2.68\ \mathrm{mm}$, which represents a rather small distance. This means that even if the greatest dimension of the vessel containing the reacting solution were no more than 1 cm, the time required for thoroughly efficient mixing achieved solely by molecular diffusion would be a few times greater than $l^2/6D$, that is, a few times 2.3 hours. It follows that, for chemical reactions in solution that are to be studied over a period of a few minutes, complete mixing must be achieved at the start if the results obtained are to have any significance whatever.

Another matter of some importance in designing a study of the kinetics of a reaction is that the stoichiometry of the latter is well established. The issue is, to what extent do side-reactions occur and is their extent dependent on experimental conditions, such as temperature?

This question is all the more important where the progress of the reaction is followed by monitoring the build-up of one of the products. Clearly, if side-reactions occur, then the concentration of a main product may never attain its theoretical value. If the concentration of the reactant, which is the parameter required in all calculations in regard to reaction order and rate constant, were to be calculated from its known initial value less that extent of reaction implied by the build-up of this reaction product, then errors may be made which may potentially be serious in the later stages of the reaction. If the side-reaction occurred even to the extent of a few percent this could lead to a very large relative error in the concentration of the reactant, especially after the reaction is 90% complete.

2.2 The evolution of techniques for monitoring reaction progress

Reactions in solution make up a very large proportion of the chemical reactions whose kinetics have been subjected to detailed investigation. The classical method of following the progress of such a reaction is to withdraw aliquots at known times, chill or treat them in some other appropriate way to arrest the reaction and then determine the concentration of one of the reactants or of one of the product species by chemical analysis.

For example, for the alkaline hydrolysis reaction,

$$ClCH_2CO_2^- + OH^- = HOCH_2CO_2^- + Cl^- \qquad (2.2)$$

which occurs at a reasonable rate while in an aqueous solution under reflux, aliquots of $5\,cm^3$ might be removed at 10 minute intervals and the reaction might be arrested by pouring each into a flask containing ice/water, and the concentration of hydroxide ion in each might readily be measured by titrating the aliquot with standard dilute acid using an indicator.

A routine such as this can easily lead to reliable results for a reaction that is slow or very slow, with a half-life in excess of an hour. Even then, the question arises as to the extent to which the reaction is disturbed by the periodic withdrawal of an aliquot of the reaction mixture. Is any foreign substance introduced as a consequence, even in apparently trivial amounts? Is the steady temperature of the reaction mixture upset by the intrusion of the apparatus for sample removal? This latter point becomes the more serious the more frequently aliquots have to be withdrawn, and is thus of greater importance with faster reactions.

Another aspect of the above procedure is the issue of the time elapsed since the commencement of the reaction to which the measurement of the reactant concentration in the aliquot properly relates. Clearly the removal and quenching of an aliquot is not an instantaneous process and the actual temperature decrease is not the ideal vertical drop. A good approximation for the operative time is the point at which half the aliquot has been delivered into the ice/water, but the estimated time for the sampling is liable to an error of several seconds. This underlines the limitation of such a method to slow and very slow reactions.

For some fraction of reactions in solution it is possible to add additional reagents to the reacting mixture so that a perceptible colour change occurs when a known extent of reaction has taken place. For example, if studying the hydrolysis reaction,

$$RCl + H_2O = ROH + HCl \tag{2.3}$$

the addition of a small amount of NaOH and an acid–base indicator results in a colour change when sufficient HCl has been produced to neutralise that amount of NaOH that had been added.

However, for the timing of this colour change to be significant it is essential that the added reagents do not in any way interfere with the primary reaction whose kinetics are to be studied. In regard to the above example, this means that the method is a valid one only if the hydrolysis of RCl is strictly independent of pH. This is one of the reasons why the method, often referred to as the clock method, is far from being a general one for monitoring reaction progress.

The first known example of a detailed study of the progress of a chemical reaction was the work of Wilhelmy in 1850 on the hydrolysis of sucrose in acid solution using a polarimeter. Coincidentally, this study also represented the first use of a physical property of the reacting mixture to monitor the progress of the reaction, an approach which is the essence of any modern technique used in the experimental study of chemical kinetics.

As a method, polarimetry has obvious advantages over those referred to above. All problems in regard to sampling are obviated. From each experiment, any number of readings may be taken, each leading to the concentration of the reactant species at a known time, and the making of such measurements causes no interference to the chemical reaction under study. These represent advantages which are in general realised when measurements of a physical property are used to monitor reaction progress.

The formal equation for the chemical reaction studied by Wilhelmy is:

$$\text{sucrose} + H_2O = \text{glucose} + \text{fructose} \tag{2.4}$$

Since sucrose, glucose and fructose are all optically active substances, a value of the angle of rotation of polarised light is a function of the concentration of all three of these species. Whereas sucrose ($\alpha = +66.5°$) and glucose ($\alpha = +52.7°$) are both dextrarotatory, fructose ($\alpha = -92.4°$) is strongly laevorotatory with the result that, as the hydrolysis reaction proceeds, the angle of rotation becomes less positive, passes through zero and becomes negative. For this reason, this chemical process has been widely referred to as "the inversion of sucrose". Calculation shows that the zero angle of rotation is attained at the point where the reaction has gone *ca* 86% of the way to completion. More generally, if one knows the initial angle of rotation due to sucrose alone, from the polarimeter reading at any subsequent time, the concentration of sucrose then remaining may be calculated unambiguously.

While polarimetry was the technique used in Wilhelmy's classic experiments in 1850, its application to monitoring reaction progress has been extremely limited since only a tiny minority of chemical reactions involve an optically active material. Studies from around 1870 onwards of the kinetics of chemical reactions were for the most part carried out by periodically withdrawing aliquots for chemical analysis. As laboratory instrumentation has become increasingly available since the 1930s, an increasing proportion of chemical kinetic studies have been carried out by monitoring a physical property of the reaction mixture. The most important of these techniques are now considered in more detail.

2.3 The application of spectrophotometry to chemical kinetics

The simplest examples of the use of spectrophotometry to monitor the progress of a chemical reaction would be in cases such as the following:

(i) One reactant species absorbs moderately or strongly in the visible or the ultra-violet, whereas all the product species are transparent over this wavelength range.

(ii) One product species absorbs moderately or strongly in the visible or the ultra-violet, whereas all the reactant species are transparent over this wavelength range.

The absorption of monochromatic light by a single species, of concentration c, is summarised by the Beer–Lambert law,

$$\log_{10}\left(\frac{I_o}{I_t}\right) = \varepsilon c l = A, \text{ the absorbance} \tag{2.5}$$

where I_o is the intensity of the incident and I_t that of the transmitted beam of light, l is the path length of the beam through the solution and ε is the molar decadic absorption coefficient at that wavelength. Thus the value of the absorbance, A, is proportional to the concentration of, in case (i), a reactant species or, in case (ii), a reaction product.

The former situation is exemplified by the reaction between permanganate ion and oxalic acid:

$$2MnO_4^- + 5C_2O_4H_2 + 6H^+ = 2Mn^{2+} + 10CO_2 + 8H_2O \tag{2.6}$$

Whereas the permanganate ion absorbs moderately strongly in the visible, with its wavelength of maximum absorption (λ_{max}) at 532 nm, no other species absorbs at wavelengths in excess of 300 nm. Thus the progress of the reaction is readily monitored by observing the decrease in the absorbance due to the permanganate ion.

An example of case (ii) is provided by the reaction,

$$H_2O_2 + 2I^- + 2H^+ = I_2 + 2H_2O \tag{2.7}$$

where the reactants are all transparent except in the far ultra-violet, but iodine absorbs in the near UV. However, in hydroxylic solvents the association process occurs,

$$I_2 + I^- \rightleftharpoons I_3^- \tag{2.8}$$

and the ion I_3^- is easily estimated spectrophotometrically, since it has a high molar absorption coefficient and a λ_{max} at 352 nm. But the absorbance at this wavelength will be proportional to the yield of reaction product only if the I^- concentration is sufficiently high that the fraction of the product iodine present as I_3^- is constant.

The usual wavelength range of a modern UV spectrophotometer is 190–900 nm, and quite often one finds that there are both reactant and product species absorbing light within this range. It is then desirable that the λ_{max} values of the reactant and the product are well separated, but in principle the progress of the reaction could be followed even if they coincide, provided that the respective absorption coefficients are substantially different.

One beneficial by-product of having appreciable absorption by both the reactant and the product of the reaction, but at different wavelengths of λ_{max}, is that then there should be a wavelength lying between these positions, known as an isosbestic point, at which the absorbance should remain constant throughout the reaction. However, this will occur only if the reaction is a simple inter-conversion with no intermediate product and an absence of any further reaction. In some instances, where both the reactant and the product have several absorption bands, there are a number of isosbestic points and the pattern expected for a simple inter-conversion is then too elaborate to be a mere coincidence.

If the reaction stoichiometry is represented by the equation

$$B = C \tag{2.9}$$

and the absorbance, A, measured at some suitable wavelength is a linear function of the concentration of both these species, that is,

$$A = \beta[B] + \gamma[C] + \alpha \tag{2.10}$$

then initially, when $[B] = b$ and $[C] = 0$:

$$A_o = \beta b + \alpha \tag{2.11}$$

When the reaction is complete, so that $[B] = 0$ and $[C] = b$, then we will have,

$$A_\infty = \gamma b + \alpha \tag{2.12}$$

but in general, at time t when the concentration of B has fallen from b to $(b - x)$, we have,

$$A_t = \beta(b - x) + \gamma x + \alpha \tag{2.13}$$

From equations (2.11) to (2.13), the following equations may be deduced:

$$A_o - A_\infty = (\beta - \gamma)b \tag{2.14}$$

$$A_o - A_t = (\beta - \gamma)x \tag{2.15}$$

$$A_t - A_\infty = (\beta - \gamma)(b - x) \tag{2.16}$$

If reaction (2.9) were kinetically first order, then the appropriate integrated rate equation would be equation (1.16)

$$\ln\left(\frac{b}{b - x}\right) = k_1 t \tag{1.16}$$

Substituting from equations (2.14) and (2.16) we have,

$$\ln\left(\frac{A_\infty - A_o}{A_\infty - A_t}\right) = k_1 t \tag{2.17}$$

which indicates that the rate constant for reaction (2.9) may be found from the slope of the plot of the left hand side against time or from minus one times the slope of the plot* of $\ln(A_\infty - A_t)$ against time. Clearly, the result of using equation (2.17) will be all the more reliable the greater the relative change in the absorbance during the course of the reaction, that is, the more β/γ differs from unity.

The use of equation (2.17) requires that the value for A_∞ be available. After ten half-lives, the reaction will have gone 99.9% of the way to completion, and a reading taken thereafter is late enough to represent completion of the reaction. However, in some cases it may prove to be unreliable because, for example, of a further reaction of the product species C. For the purposes of using in equation (2.17), errors in A_∞ become increasingly important the higher the value of t, since as $(A_\infty - A_t)$ becomes progressively smaller, the error in $\ln(A_\infty - A_t)$ correspondingly increases. This effect is illustrated in Table 2.1.

If the reaction being studied were to follow second order kinetics, then equation (1.20) would apply

$$\frac{1}{b - x} - \frac{1}{b} = k_2 t \tag{1.20}$$

Substituting from equations (2.14) and (2.16) above, this gives:

$$k_2 t = \frac{1}{b} \cdot \frac{x}{b - x} = \frac{1}{b} \cdot \left(\frac{A_t - A_o}{A_\infty - A_t}\right) \tag{2.18}$$

*This assumes that the absorbance rises as the reaction proceeds, so that A_∞ is greater than A_t. If the reverse were true, then the function to be plotted is $\ln(A_t - A_\infty)$.

Table 2.1 Illustration of how errors of constant magnitude in ΔA will give rise to increasing errors in $\ln(\Delta A)$ as ΔA decreases.

t	ΔA	$\ln(\Delta A)$
0	1.450 ± 0.02	0.37 ± 0.01
2	1.118 ± 0.02	0.11 ± 0.02
4	0.862 ± 0.02	-0.15 ± 0.02
6	0.665 ± 0.02	-0.41 ± 0.03
8	0.513 ± 0.02	-0.67 ± 0.04
10	0.385 ± 0.02	-0.93 ± 0.05
12	0.305 ± 0.02	-1.19 ± 0.07
14	0.235 ± 0.02	-1.45 ± 0.09
16	0.181 ± 0.02	-1.71 ± 0.11
18	0.140 ± 0.02	-1.97 ± 0.15
20	0.108 ± 0.02	-2.23 ± 0.25
22	0.083 ± 0.02	-2.75 ± 0.32
24	0.064 ± 0.02	-3.02 ± 0.43

Thus, for a second order reaction, $\{(A_t - A_0)/(A_\infty - A_t)\}$ should be a linear function of time, with the slope equal to k_2b. However, a reliable value for A_∞ is now much more elusive, in that the time required for 99.9% reaction is now a thousand times that required for the first half of the reaction to occur. In studying a second order reaction it is preferable to use conditions where the reaction is pseudo-first order, provided that the reactant present in excess is not the light-absorbing species.

2.4 Electrical conductivity and dilatometry

Where the reaction to be studied takes place in a highly polar solvent and involves ionic species, it may be appropriate to follow its progress conductimetrically. The alkaline hydrolysis of an ester serves to illustrate the application of this technique.

$$RCO_2Et \quad + OH^- \quad = \quad RCO_2^- \quad + EtOH \tag{2.19}$$

As reaction proceeds, the electrical conductivity of the solution decreases because the conductance of the ion RCO_2^- formed is appreciably less than that of the reacting hydroxide ion.

While equation (2.19) adequately represents the stoichiometry of the reaction, there must inevitably be present a counter-ion, such as Na^+, whose concentration will remain constant at the initial concentration of OH^-. While the Na^+ ion will necessarily add to the specific conductivity of the solution, this contribution will be small in relation to that made initially by the OH^- ion.

Since, in this system, the specific conductance κ is given by:

$$\kappa = \frac{1}{10^3}\Sigma(c_i\lambda_i) \tag{2.20}$$

where c_i denotes the concentration in mol dm^{-3} and λ_i the ionic conductance in S cm^2 mol^{-1} of the *i*th ion present in the solution. One may therefore express the specific conductance, κ, in terms of the concentrations of the ions OH^-, Na^+ and RCO_2^-, in equations totally analogous to those for the absorbance A in the previous section, that is equations (2.11) to (2.13).

In studying the kinetics of a reaction such as (2.19) by a conductimetric technique, it is desirable to operate under pseudo-first order conditions, since under second order conditions it is very difficult to obtain a reliable value for κ_∞. However it is important that the ester and not the hydroxide is present in large excess, so that the conductivity undergoes a substantial relative change in the course of the reaction, thus leading to a more reliable value for the rate constant. The relevant equation, totally analogous to equation (2.17) above, is:

$$\ln\left(\frac{\kappa_0 - \kappa_\infty}{\kappa_t - \kappa_\infty}\right) = k't \tag{2.21}$$

where k' represents the phenomenological first order rate constant.

Dilatometry, the measurement of small changes in the volume of a solution as reaction occurs, is a technique that has been widely used, particularly to monitor the progress of addition polymerisation processes. Its applicability rests on the fact that, in solution, the partial molar volume of the monomer is not identical to the partial molar volume per monomer unit of the polymer. However, such differences in the partial molar volumes of reactants and products are often so small that careful measurements are necessary in order to use them for kinetic purposes.

In practical terms, the solution is contained within a vessel with a capillary tube at the top, as shown in Figure 2.1. The movement of the meniscus is followed using a travelling microscope. In such a system, thermostatting is doubly necessary, since temperature affects both the density of the solution and the reaction rate.

2.5 Techniques for the gas phase

The classical method of monitoring a reaction in the gas phase is to follow the change in the total pressure. Thus in the pyrolysis of dimethyl ether,

$$CH_3OCH_3 = CH_4 + CO + H_2 \tag{2.22}$$

where one molecule reacts to yield three product molecules, the total pressure is trebled in the course of the reaction. In the catalytic hydrogenation of ethene on the surface of a transition metal,

$$C_2H_4 + H_2 = C_2H_6 \tag{2.23}$$

the total pressure decreases during the course of the reaction, by a factor which depends on the relative initial amounts of the two reactants.

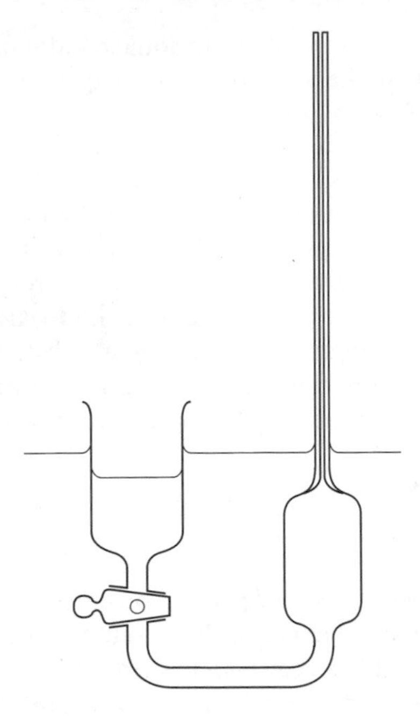

Fig. 2.1 A dilatometer, immersed in a thermostat tank, where the movement of the meniscus in the capillary may be followed using a travelling microscope.

However, while this technique was used in early investigations of reactions such as (2.22) and (2.23), it suffers from the limitation that it is useful only if the reaction causes a net change in the number of molecules present. There is the additional point that the device used to measure pressure may require to be kept at a temperature which differs from that required to bring about the reaction which is to be monitored.

A more satisfactory technique for reactions in the gas phase is mass spectrometry. The operational details of mass spectrometers vary from one device to another, but in all cases there are two distinguishable phases. Firstly, there is the ionisation stage, where ions characteristic of the substance are produced and then in the next stage these ions are separated on the basis of their differing mass-to-charge ratios so that they can each be detected and measured.

It is preferable if, in studying a reaction, the concentration of each substance is monitored using an ion with an m/z value produced uniquely by that substance. Thus reaction (2.22) could be followed using the ions at $m/z = 46$ for dimethyl ether, 16 for methane and 28 for CO. In reaction (2.23), the peak at $m/z = 30$ is indicative only of ethane, but there is no ion derived from ethene that cannot also be derived from ethane. However, careful calibrations using ethane will suffice to determine the appropriate correction to the $m/z = 28$ peak for the contribution from ethane, and the remainder of this peak will be a measure of the concentration of ethene.

The mass spectrometer is, of course, the instrument required in assays involving stable isotopes. Some examples of these applications will be referred to in subsequent chapters.

2.6 Modern experimental techniques

In utilising the raw experimental data on a chemical reaction, a great many repetitive calculations need to be performed. The computer is ideally equipped for such a task since it can readily perform such calculations accurately and also incredibly quickly.

The principles involved are readily illustrated by reference to a reaction in solution that may be studied spectrophotometrically. Under conditions where the reaction is slow, with a half-life of at least a few minutes, it would be appropriate to start the reaction by mixing two solutions, both previously maintained for some time at the temperature selected for the reaction, filling a spectrophotometer cell with the resulting solution, and setting this into the cell-holder, thermostatted at the chosen temperature, in the cell compartment of the spectrophotometer. The output of the spectrophotometer, as the absorbance of the solution at the chosen monitoring wavelength, is then fed to the computer where the value of the absorbance, A, at chosen intervals of time is recorded into the memory.

Where the reaction occurs more quickly than this, a more appropriate approach is to use a stopped-flow apparatus. This may be thought of as a technique which evolved from the flow methods pioneered by Hartridge and Roughton in the 1920s, but modified so that small amounts of solution suffice instead of the many litres needed in a continuous flow experiment. But in a sense it is simply a device to achieve rapid and efficient mixing of two solutions, so that the ensuing reaction may be monitored, commencing a few milli-seconds after the mixing has taken place.

A diagram of a typical stopped-flow instrument is shown in Figure 2.2. From two reservoirs, the syringes S_1 and S_2 are filled with their respective solutions. To start the reaction, these solutions are forced out of their syringes by concerted pressure on the plungers, through the mixing chamber, M, and along the observation tube to the collector syringe, S_3. When the plunger of the latter reaches its stop, this both arrests the flow of the mixing solutions and triggers the recorder. From that point onwards, the amount of light, of the wavelength setting of the monochromator, transmitted through the mixed solution at a point, P, just downstream of the mixing chamber, will undergo changes as the reaction proceeds in the now static mixed solution. A photomultiplier unit converts this transmitted light into an electric current, a grid resistor generates a voltage proportional to the light intensity and, using an analogue-to-digital (A/D) converter, this signal is fed to the computer, which samples at pre-determined intervals over an appropriate time interval.

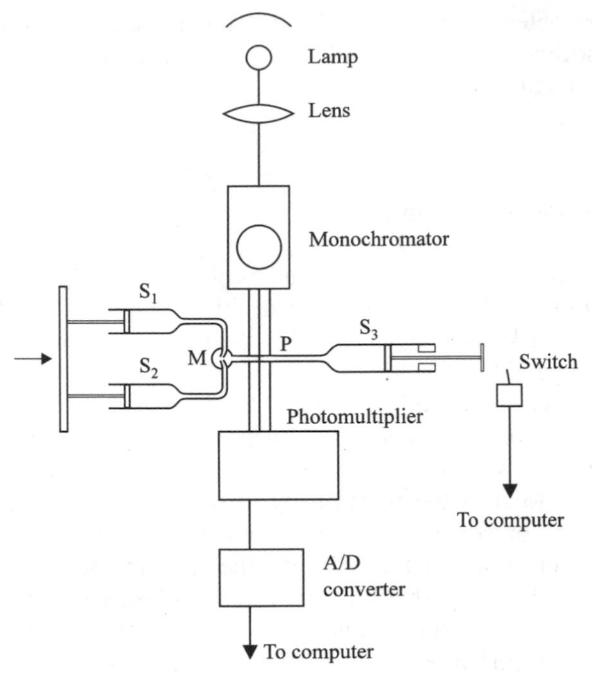

Fig. 2.2 A schematic diagram of a stopped flow spectrometer, with syringes S_1 and S_2 containing the reactant solutions. M designates the mixing chamber, and P the point at which the reaction is monitored using a monochromatic light beam, which is detected by a photomultiplier and the output fed to the computer.

There are various ways in which kinetic data of this type may be processed in order to evaluate the pseudo-first order rate constant. While the computer can be programmed to do this automatically, without any operator participation, there are merits in having the various steps carried out under the operator's direction. One reason for this is that, whether on account of noise in the photomultiplier circuit or of malfunction by the xenon lamp, it is possible that a few of the sampled points may be appreciably in error and so need to be edited out, which is a task best done under the operator's control. Another factor is that the operator's involvement in directing the calculation must aid awareness of any unusual and perhaps revealing features of the data that have been recorded. In this way it is hoped that any experimental problems will be detected at an early stage.

The first step in the treatment of the data is to convert the intensity of the transmitted light, I_t, into an absorbance:

$$A_t = \log_{10}(I_o/I_t) \tag{2.5}$$

While an error in the I_o value would lead to an incorrect value for the absorbance, A, the effect of so doing is not so serious. Suppose instead of the correct value, I_o, one uses a value I^*:

$$\begin{aligned}\log_{10}(I^*/I_t) &= \log_{10}(I^*/I_o) + \log_{10}(I_o/I_t) \\ &= D + A_t\end{aligned} \tag{2.24}$$

As equation (2.24) demonstrates, the consequence is that every value for the absorbance is in error by the constant amount, D, so that the parameters $(A_\infty - A_o)$ and $(A_\infty - A_t)$ are unaffected by the use of the incorrect datum for I_o.

2.7 The evaluation of the rate constant

Assuming that the reaction is being studied spectrophotometrically under conditions where it follows first order kinetics, the first order (or pseudo-first order) rate constant is easily obtained from the plot of $\ln(A_\infty - A_t)$ against t. However, in some instances there are difficulties in obtaining a reliable value for A_∞, which is used in calculating the ordinate of every point of this graph. This situation may arise if the light-absorbing reaction product undergoes any further reaction. Even a very slight extent of further reaction, over the several half-lives normally allowed for the reaction to be essentially complete, would be sufficient to make the absorbance reading unacceptable as a value for A_∞.

A simple procedure has been proposed by Guggenheim for solving this difficulty, and is applicable provided that readings of A have been taken at regular time intervals.

Starting from equation (2.17),

$$\ln\left(\frac{A_\infty - A_o}{A_\infty - A_t}\right) = k_1 t \tag{2.17}$$

and assuming that the product absorbs more strongly so that the absorbance rises as the reaction proceeds, by taking antilogarithms we have:

$$(A_\infty - A_t) = (A_\infty - A_o)\mathrm{e}^{-k_1 t} \tag{2.25}$$

If Δ denotes a fixed interval of time, then we may write:

$$(A_\infty - A_{t+\Delta}) = (A_\infty - A_o)\mathrm{e}^{-k_1(t+\Delta)} \tag{2.26}$$

By subtracting equation (2.26) from equation (2.25) we obtain:

$$A_{t+\Delta} - A_t = (A_\infty - A_o)(1 - \mathrm{e}^{-k_1\Delta})e^{-k_1 t} \tag{2.27}$$

Taking natural logarithms, this leads to:

$$\ln(A_{t+\Delta} - A_t) = \ln\{(A_\infty - A_o)(1 - \mathrm{e}^{-k_1\Delta})\} - k_1 t \tag{2.28}$$

Thus $\ln(A_{t+\Delta} - A_t)$ is a linear function of t, with the slope equal to $-k_1$. However, with increasing time, the left hand side of this equation becomes increasingly liable to error so that the points become progressively less reliable.

This is demonstrated in Table 2.1 in which the data for the A values were obtained by starting with an exponential decay, over more than four half-lives, to which random errors were assigned, on the basis that errors of a certain magnitude are equally probable at any value of A. As is apparent, the probable error in $\ln(A_{t+\Delta} - A_t)$ tends to increase with increasing values of t.

The usual way to determine the slope of the best straight line through a group of n points, (x_i, y_i) for $1 \leq i \leq n$, is to apply the familiar least squares regression formula:

$$m = \frac{n\Sigma(x_i y_i) - \Sigma x_i \Sigma y_i}{n\Sigma x_i^2 - (\Sigma x_i)^2} \tag{2.29}$$

However, this formula is derived on the explicit assumption that all points are subject to the same range of errors in y. When applied to a set of points such as those referred to in the previous paragraph, there is a real danger that the line will be unduly influenced by erratic points at the higher values of t. Quite commonly, in order to obviate this danger, all points after a chosen value of t are edited out, but this procedure is far from being ideal.

In the present circumstances, it is more reasonable to replace the assumption that the best straight line is the one which minimises the squares in the residuals in y, with the aim to minimise the sum, over all the points, of the ratio of the residual to the probable error in y. If the characteristic point has co-ordinates x_i and y_i and the probable error in y is γ_i, then when this aim is achieved the slope and intercept are given by:

$$m_o = \frac{\Sigma\left(\frac{x_i y_i}{\gamma_i^2}\right)\Sigma\left(\frac{1}{\gamma_i^2}\right) - \Sigma\left(\frac{x_i}{\gamma_i^2}\right)\Sigma\left(\frac{y_i}{\gamma_i^2}\right)}{\Sigma\left(\frac{x_i^2}{\gamma_i^2}\right)\Sigma\left(\frac{1}{\gamma_i^2}\right) - \left\{\Sigma\left(\frac{x_i}{\gamma_i^2}\right)\right\}^2} \tag{2.30}$$

$$c_o = \frac{\Sigma\left(\frac{x_i^2}{\gamma_i^2}\right)\Sigma\left(\frac{y_i}{\gamma_i^2}\right) - \Sigma\left(\frac{x_i}{\gamma_i^2}\right)\Sigma\left(\frac{x_i y_i}{\gamma_i^2}\right)}{\Sigma\left(\frac{x_i^2}{\gamma_i^2}\right)\Sigma\left(\frac{1}{\gamma_i^2}\right) - \left\{\Sigma\left(\frac{x_i}{\gamma_i^2}\right)\right\}^2} \tag{2.31}$$

In order to apply equations (2.30) and (2.31) it is necessary to quantify the probable error in each point and to use these data in the calculation. However, in the case of plots from first order or pseudo-first order reactions, of $\ln(A_\infty - A_t)$ against time (equation 2.17) or of $\ln(A_{t+\Delta} - A_t)$ against time (equation 2.28), the situation is a little simpler. Each absorbance value may be assumed liable to the same experimental error and we shall represent the possible range of the absorbance difference as $(\Delta A \pm \delta)$. The quantity plotted on the graph is $\ln(\Delta A \pm \delta)$:

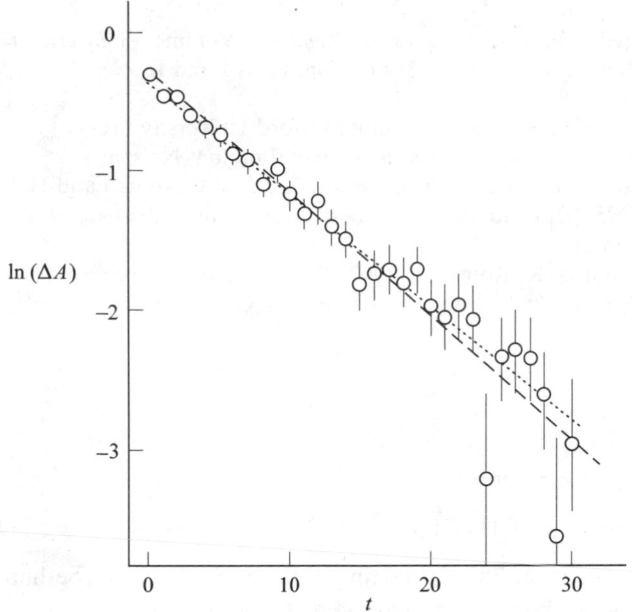

Fig. 2.3 A plot of $\ln(A_{t+\Delta} - A_t)$ against time, using data to which random errors have been introduced. The linear least squares line, from equation (2.29) is represented (– – –) and the line from equation (2.33) is shown as (- - - - -).

$$\begin{aligned} \ln(\Delta A \pm \delta) &= \ln(\Delta A) + \ln(1 \pm \delta/\Delta A) \\ &\approx \ln(\Delta A) \pm \delta/\Delta A \end{aligned} \tag{2.32}$$

Thus the probable error in the logarithm of ΔA is inversely proportional to ΔA, so that it is not necessary to estimate the probable error of each point individually.

This means that the slope of the optimum straight line of the plot of ΔA (as the ordinate) against t will be given by the formula:

$$m_1 = \frac{\Sigma(\Delta A_i^2)\Sigma(\Delta A_i^2 \cdot t_i \ln \Delta A_i) - \Sigma(\Delta A_i^2 \cdot t_i)\Sigma(\Delta A_i^2 \cdot \ln \Delta A_i)}{\Sigma(\Delta A_i^2)\Sigma(\Delta A_i^2 \cdot t_i) - \{\Sigma(\Delta A_i^2 \cdot t_i)\}^2} \tag{2.33}$$

The consequence of using this equation rather than the conventional linear least squares regression formula (equation 2.29) is illustrated in Figure 2.3. The "theoretical" equation for the decay curve, before the random errors were added, is:

$$\ln(A_{t+\Delta} - A_t) = -0.3584 - 0.0800t \tag{2.34}$$

By the conventional least squares formula, the rate constant is evaluated as 0.0828, but from equation (2.33) the value obtained is 0.0803, which is significantly more accurate.

Suggested reading

Hammes, G. G. (ed.), 1974. *Techniques of Chemistry*, Volume VI *Investigations of rates and mechanisms of reactions*, 3rd edition, Parts I and II. New York: Wiley-Interscience.

Bradley, J. N. 1975. *Fast Reactions.* London: Oxford University Press.

Swinbourne, E. S., 1971. *Analysis of Kinetic Data.* London: Nelson.

Francis, P. G., 1984. *Mathematics for Chemists.* London: Chapman and Hall.

Logan, S. R., 1995. How to draw the best straight line. *Journal of Chemical Education,* **72**, in press.

Bevington, P. R., and D. K. Robinson, 1992. *Data Reduction and Error Analysis for the Physical Sciences*, 2nd edition. New York: McGraw-Hill.

Problems

2.1 For the gas phase reaction,

$$CH_2Cl.CH_2Cl \quad = \quad CH_2{:}CHCl + HCl$$

in a kinetic experiment at 780 K, starting with pure 1,2-dichloroethane, the total pressure p was observed to increase as follows:

t/s	0	45	90	135	180	225	270
p/torr	150	176.8	198.8	216.8	231.7	243.9	253.9

Show that the reaction follows first order kinetics and evaluate the rate constant from these data.

2.2 The reaction of 1,1-diphenyl-2-picrylhydrazyl (DPPH) with iron(II) ion in water/ethanol solution was followed spectrophotometrically. In an experiment at 293 K, the absorbance A (due to DPPH), was found to decrease with time as follows:

t/s	0	17	37	37	76	94	114	138	157	178	204	∞
A	0.54	0.49	0.44	0.40	0.36	0.33	0.30	0.27	0.25	0.23	0.21	0.11

Given that, at $t = 0$, the reactant concentrations were [DPPH] $= 1.14 \times 10^{-5}$ mol dm^{-3} and $[Fe^{2+}] = 2.75 \times 10^{-4}$ mol dm^{-3}, evaluate the second order rate constant for this reaction.

2.3 Assuming that, in Problem 1.2, each value for the concentration of R· is liable to the same absolute error, determine the rate constant for the decay of R· using equation (2.30).

2.4 The main products from the thermal decomposition of tetraethyl lead are metallic lead and butane/isobutane, with some ethane and ethene. In an experiment at 548 K, starting with a sample of pure tetraethyl lead, the gas pressure was observed to rise as follows:

t/s	0	60	110	155	231	362	564	∞
P/torr	19.0	21.5	23.3	24.9	27.1	29.9	33.1	38.8

Assuming that the stoichiometry of the reaction remains constant through the decomposition, show that the reaction follows first order kinetics and evaluate the rate constant. (Data from Leermakers, J. A., *J. Am. Chem. Soc.*, 1933, **55**, 4508–18.)

2.5 The alkaline hydrolysis of ethylene carbonate,

$$\begin{matrix} CH_2-O \\ | \quad\quad | \\ | \quad C{=}O \\ | \quad\quad | \\ CH_2-O \end{matrix} + 2OH^- = \begin{matrix} CH_2OH \\ | \\ CH_2OH \end{matrix} + CO_3^{2-}$$

was monitored by a conductivity method. In one experiment at 298 K, with initial concentrations $[OH^-]_0 = 0.00514\ \text{mol dm}^{-3}$ and $[\text{ester}]_0 = 0.00081\ \text{mol dm}^{-3}$, the oscillator current I, which is a measure of the specific conductivity of the solution, was found to vary with time as follows:

t/s	0	7.5	15	22.5	30	37.5
I	80.6	77.7	75.0	72.3	70.0	67.5

t/s	45	60	67.5	75	90	∞
I	65.5	62.0	60.5	59.0	56.5	37.6

Deduce the rate constant for the reaction of the first hydroxide ion with the ester, assuming that the reaction of the second OH^- follows closely. (Data from Saadi, A. H. and W. H. Lee, *J. Chem Soc., B*, 1966, 1–4.)

3 Reaction mechanism and reaction order

While some chemical reactions occur in a single step, a great many involve a number of individual steps. The manner in which these reaction steps are linked together dictates the way in which a reactant is perceived to be consumed and a reaction product to be formed. It is for this reason that the study of reaction rates may elucidate the reaction mechanism.

In this chapter we consider some simple reaction mechanisms and the consequent kinetic behaviour. We also look at isotopic exchange reactions, which display certain notable kinetic features regardless of the mechanisms involved.

3.1 An elementary reaction and the molecularity

The simplest way in which we may conceive of a chemical reaction taking place is that the individual molecules corresponding to the left hand side of the stoichiometric equation for the reaction should meet and, as a consequence of the rapid re-location of some of the bonding electrons, yield the species written on the right hand side of the equation. Thus, in one single step, the reaction will have taken place.

The term employed to denote a process of the above type is an *elementary* reaction. These reactions are sub-divided on the basis of the number of molecules, including ions and free radicals, that take part in the reaction process. In many cases, two reacting species are involved and the reaction is described as a bimolecular one. Occasionally, just one molecule is involved and the reaction is described as unimolecular, and rarely there are three species involved in what is therefore a termolecular process. The molecularity is thus a positive integer and values greater than three are unknown.

As an example of a unimolecular reaction we may cite the isomerisation of cyclopropane, in the gas phase at elevated temperatures, to propene.

$$\overset{\displaystyle CH_2}{CH_2\!\!-\!\!-\!\!CH_2} \quad \longrightarrow \quad CH_3CH:CH_2 \qquad (3.1)$$

All nine atoms of the product molecule are the same as formerly made up the reactant molecule.

To exemplify a bimolecular reaction we may cite the reaction of ethyl ethanoate with hydroxide ion:

$$CH_3CO_2C_2H_5 + OH^- \rightarrow CH_3CO_2^- + C_2H_5OH \qquad (3.2)$$

This reaction occurs in solution and although the role of solvent molecules may not quite be that of totally detached spectators, such solvent molecules are not included among the molecules or ions that are counted as participants.

Over the decades, there has been much confusion over the inter-relationship of the reaction order and the molecularity of a reaction. In part this is perhaps because, long after Ostwald had first introduced the term "reaction order", other equally illustrious scientists continued to refer to what was clearly the same parameter as the "molecularity". In this way, the seeds of confusion were sown by the pioneers. But perhaps the persistence of confusion has owed more to the fact that statements of the concept of molecularity have often omitted to specify that this is a parameter relating to an elementary reaction, or at least to an individual step in a reaction sequence, and has no meaning in regard to a reaction of complex mechanism.

A reaction occurring in a single step, involving precisely those species on the left hand side of the stoichiometric equation, is special in certain ways. Firstly, in recognition of its elementary status, it is conventional to write it with an arrow. In this book, the arrow is used solely for that purpose, and so an equality sign is used elsewhere merely to denote the material balance of a reaction. Equations (3.1) and (3.2) have accordingly been written with an arrow.

Also, if a reaction is thought to take place in an elementary process, then the kinetic order is readily deduced. A rearrangement that occurs in a unimolecular process will follow first order kinetics, just as does radioactive decay: the rate must be proportional to the concentration of molecules remaining to react. A bimolecular reaction will exhibit second order kinetics, since the rate must be proportional to the concentration of each of two reacting species. In the case of a reaction such as (3.2), it will be first order in respect of each of two reactants. For the combination of iodine atoms in solution,

$$I + I \rightarrow I_2 \qquad (3.3)$$

one must expect the reaction to be second order in iodine atoms, since two such species participate in the reaction step. Similarly, for a termolecular reaction one anticipates third order kinetics.

However, it is important to note that the converse to the arguments of the previous paragraph is not valid. If, for example, the reaction between A and B is found by experiment to be kinetically first order in each reactant, this does not necessarily mean that the reaction is an elementary one. While such a single-stage process is clearly consistent with the experimental facts, there may be other plausible mechanisms that would also fit these findings. For that reason, one should not presume.

There are many different kinds of reactions that are clearly not elementary processes. Two such groups are so important that they have chapters devoted specially to them: chain reactions are dealt with in Chapter 7 and càtalysed reactions in Chapters 8 and 9. In this chapter we will consider some simple examples of non-elementary or composite reactions.

3.2 Consecutive reaction processes

Reactions that occur by consecutive steps are well illustrated by the hydrolysis reactions of certain alkyl halides, where the overall equation may be written:

$$RX + H_2O = ROH + H^+ + X^- \tag{3.4}$$

In certain cases the rate of the reaction is unaffected by the pH of the solution and the process occurs by the consecutive steps,

$$RX \rightarrow R^+ + X^- \tag{3.5}$$

$$R^+ + H_2O \rightarrow ROH + H^+ \tag{3.6}$$

where the second step very closely follows the first. Chemists interested in organic reaction mechanisms have labelled this an S_N1 reaction, where the symbols stand for "unimolecular nucleophilic substitution". A more pedantic approach would be to say that the first of the proposed two steps of the reaction is a unimolecular one.

The kinetic consequence of the fact that the reaction takes place in this way is perhaps clearer in alkaline solution, where the overall process may more precisely be written,

$$RX + OH^- = ROH + X^- \tag{3.4a}$$

and under these conditions the second step becomes:

$$R^+ + OH^- \rightarrow ROH \tag{3.6a}$$

Clearly, the rate of the reaction is controlled by the first step, reaction (3.5), whose rate is proportional only to the concentration of RX. If, on the other hand, the solvolysis process detailed in equation (3.4a) were to occur in a single step, the rate of the solvolysis would then be proportional both to the concentration of RX and to that of hydroxide ion. A mechanism of this latter type has been given the label S_N2.

As a variation of the above mechanism, we may contemplate the following:

$$\left.\begin{array}{l} A \xrightarrow{1} B \\ B \xrightarrow{2} C \end{array}\right\} \tag{3.7}$$

B represents a stable but reactive molecule, formed by the unimolecular reaction of A, and which itself reacts by another unimolecular process.

Thus we have, for the reaction of A,

$$-\frac{d[A]}{dt} = k_1[A]$$

$$\text{or} \quad -\frac{d(a-x)}{dt} = k_1(a-x) \tag{3.8}$$

so that the concentration of A will decrease exponentially: $[A] = ae^{-k_1 t}$

For the concentration of B we have:

$$\frac{d[B]}{dt} = k_1(a-x) - k_2[B]$$

$$= k_1 ae^{-k_1 t} - k_2[B]$$

$$\therefore \frac{d[B]}{dt} + k_2[B] = k_1 ae^{-k_1 t} \tag{3.9}$$

Making use of the initial condition that $[B] = 0$ at $t = 0$, the solution to this differential equation (provided that $k_1 \neq k_2$) is:

$$[B] = \frac{ak_1}{k_2 - k_1}(e^{-k_1 t} - e^{-k_2 t}) \tag{3.10}$$

The concentration of C can now be evaluated from the fact that, at any time, $[A] + [B] + [C] = a$. This gives the equation:

$$[C] = a\left\{1 - \frac{k_2}{k_2 - k_1} \cdot e^{-k_1 t} - \frac{k_1}{k_1 - k_2} \cdot e^{-k_2 t}\right\} \tag{3.11}$$

The variations with time of the concentrations of the species A, B and C are represented in Figure 3.1, where k_2 was taken as $\frac{1}{3}k_1$. It may be noted in equation (3.11) that this expression is symmetrical with respect to k_1 and k_2. Thus when the values allocated to these rate constants are interchanged, as they are in Figure 3.2, the build-up of the final product C shows just the same profile as in Figure 3.1, even though the profiles of the concentrations of A and B are quite different.

Differentiating equation (3.11) with respect to time, it is seen that $d[C]/dt = 0$ at $t = 0$. This confirms what may be seen from Figures 3.1 and 3.2, that since C is formed not from a reactant, but from an initial product, the initial rate of formation of C is zero.

On the other hand, the initial rate of formation of B is finite, and the concentration attains a maximum before it commences to decline. By differentiating equation (3.10) we obtain,

$$\frac{d[B]}{dt} = \frac{ak_1}{k_2 - k_1}(k_2 e^{-k_2 t} - k_1 e^{-k_1 t}) \tag{3.12}$$

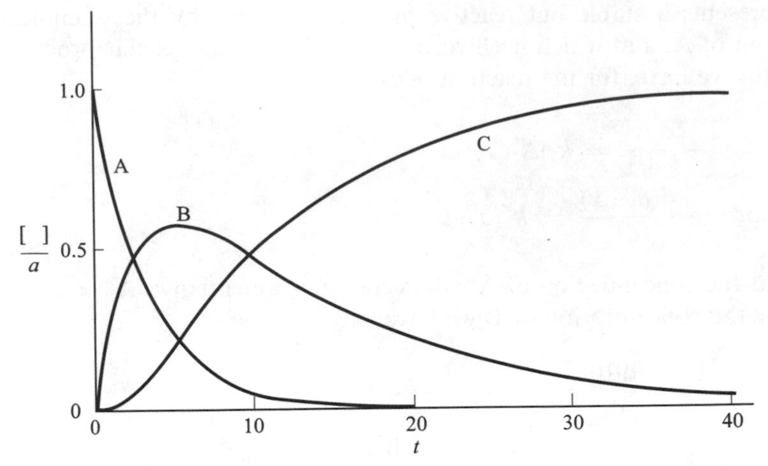

Fig. 3.1 Plot of the concentrations of A, B and C against time, in accordance with equations (3.8), (3.10) and (3.11), where k_2 was taken as $\frac{1}{3}k_1$.

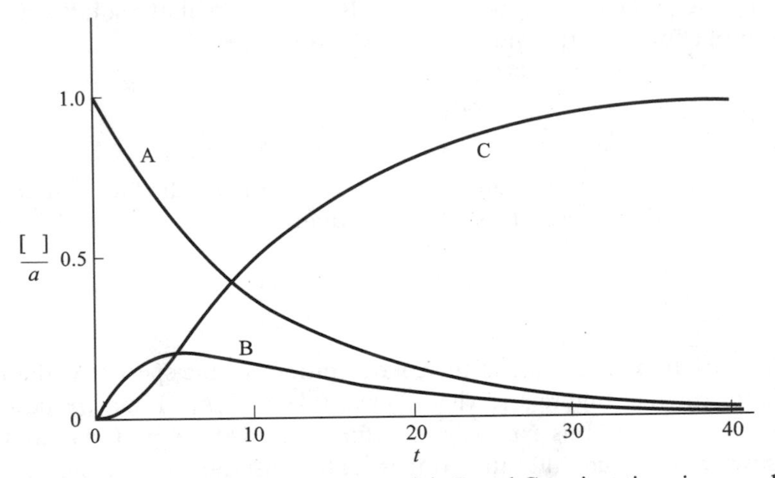

Fig. 3.2 Plot of the concentrations of A, B and C against time, in accordance with equations (3.8), (3.10) and (3.11), with k_2 equal to $3k_1$. Compared to Figure 3.1, the values of k_1 and k_2 have been inverted.

from which it may be seen that the maximum concentration of B is achieved at the time, t_m, given by:

$$t_m = \frac{1}{k_1 - k_2} \ln\left(\frac{k_1}{k_2}\right) \tag{3.13}$$

While the precise details of mechanism (3.7) will make its occurrence rare, if the formation of the initial and the final products are pseudo-first order reactions then the same kinetic scheme is applicable. For the ester of a dibasic acid, hydrolysis in the presence of a large excess of alkali would proceed in just this fashion, with similar kinetic consequences

$$\left.\begin{array}{l}
\begin{array}{c}CO_2Et\\ |\\ R'\\ |\\ CO_2Et\end{array} + OH^- \xrightarrow{1} \begin{array}{c}CO_2^-\\ |\\ R'\\ |\\ CO_2Et\end{array} + EtOH \\[2ex]
\begin{array}{c}CO_2^-\\ |\\ R'\\ |\\ CO_2Et\end{array} + OH^- \xrightarrow{2} \begin{array}{c}CO_2^-\\ |\\ R'\\ |\\ CO_2^-\end{array} + EtOH
\end{array}\right\} \qquad (3.14)$$

3.3 Formation of an intermediate complex

When two species react together, it is conceivable that rather than engaging in a bimolecular reaction, they form an intermediate complex which may either yield the reaction products or dissociate back to the reactants. The steps involved are as follows:

$$\left.\begin{array}{lcl}
A + B & \underset{-1}{\overset{1}{\rightleftarrows}} & X \\
X & \overset{2}{\rightarrow} & C + D
\end{array}\right\} \qquad (3.15)$$

This intermediate complex, X, is envisaged as differing appreciably from the initial product, B, in the consecutive reaction example in Section 3.2, in that X is much more reactive, would not normally be detectable in the reacting system, and its concentration would be negligibly small in relation to those of A and B.

To evaluate the kinetic implications of a mechanism such as this, it is extremely useful to invoke the Principle of Stationary States, which was first introduced by Chapman in 1913. This principle amounts to making the assumption that the net rate of formation of a reaction intermediate, such as X, may be put equal to zero. This permits an expression to be derived for the concentration of the reaction intermediate and thus for the rate of the steps in the reaction mechanism in which X participates.

Of course, the Principle of Stationary States cannot be strictly true. If it were, then the concentration of X, which is initially zero, would have to remain at zero. It is however an acceptable approximation because, since X is so reactive, its concentration remains very low. As we shall see in Chapter

9, there are some reactions in which the Principle is not applicable to the reactive intermediates involved.

Applying the Principle of Stationary States to mechanism (3.15) we have:

$$\frac{\mathrm{d}[\mathrm{X}]}{\mathrm{d}t} = k_1[\mathrm{A}][\mathrm{B}] - k_{-1}[\mathrm{X}] - k_2[\mathrm{X}] = 0 \tag{3.16}$$

Since the algebraic sum of these terms may be equated to zero we obtain, for the concentration of X:

$$[\mathrm{X}] = \frac{k_1[\mathrm{A}][\mathrm{B}]}{k_{-1} + k_2} \tag{3.17}$$

The rate at which the reaction products are formed is then given by :

$$\frac{\mathrm{d}[\mathrm{C}]}{\mathrm{d}t} = k_2[\mathrm{X}] = \frac{k_1 k_2[\mathrm{A}][\mathrm{B}]}{k_{-1} + k_2} \tag{3.18}$$

Thus mechanism (3.15), although not an elementary bimolecular reaction, should cause the reaction to exhibit second order kinetics, being first order with respect to both A and B. But whereas a bimolecular reaction usually gives a fairly linear Arrhenius plot, this mechanism may not. If steps –1 and 2 are affected by temperature in appreciably different ways, then mechanism (3.15) should cause the Arrhenius plot to show considerable curvature.

Alternatively, an intermediate complex may be involved in the reaction between A, B and C. Consider the mechanism :

$$\left.\begin{array}{lcl} \mathrm{A} + \mathrm{B} & \underset{-1}{\overset{1}{\rightleftarrows}} & \mathrm{X} \\ \mathrm{X} + \mathrm{C} & \overset{2}{\rightarrow} & \mathrm{P} + \mathrm{Q} \end{array}\right\} \tag{3.19}$$

Applying the Principle of Stationary States, we obtain:

$$\frac{\mathrm{d}[\mathrm{X}]}{\mathrm{d}t} = k_1[\mathrm{A}][\mathrm{B}] - k_{-1}[\mathrm{X}] - k_2[\mathrm{X}][\mathrm{C}] = 0 \tag{3.20}$$

$$\therefore\ [\mathrm{X}] = \frac{k_1[\mathrm{A}][\mathrm{B}]}{k_{-1} + k_2[\mathrm{C}]} \tag{3.21}$$

$$\therefore\ \frac{\mathrm{d}[\mathrm{P}]}{\mathrm{d}t} = k_2[\mathrm{X}][\mathrm{C}] = \frac{k_1 k_2[\mathrm{A}][\mathrm{B}][\mathrm{C}]}{k_{-1} + k_2[\mathrm{C}]} \tag{3.22}$$

Alternative extreme situations are possible in relation to equation (3.22). We may conceive of either of the two terms in the denominator being predominant to the extent that the other may be neglected. If the vast majority of intermediates X, formed in step 1, react by step 2 rather than by step –1, then the rate of step –1 is clearly much less than that of step 2. That is, $k_{-1}[\mathrm{X}] \ll k_2[\mathrm{X}][\mathrm{C}]$. Consequently, the denominator of equation (3.22) may be simplified:

$$\frac{d[P]}{dt} = \frac{k_1 k_2[A][B][C]}{k_2[C]} = k_1[A][B] \quad (3.22a)$$

The other extreme possibility is, of course, that $k_{-1}[X] \gg k_2[X][C]$ and that the vast majority of molecules X formed in step 1 react by step –1. This leads to the other limiting possibility for equation (3.21), namely :

$$\frac{d[P]}{dt} = \frac{k_1 k_2[A][B][C]}{k_{-1}} \quad (3.22b)$$

Whereas equation (3.22a) predicts that the reaction will exhibit second order kinetics and show no dependence on the concentration of C, according to equation (3.22b) the same reaction, occurring by the same mechanism, should show third order kinetics. This situation underlines the point made in the penultimate paragraph of Section 3.1.

The apparent paradox of the differing reaction orders is readily explained. Where step -1 predominates over step 2, this means that steps 1 and -1 make up an equilibrium that is scarcely perturbed by step 2. The concentration of X is then essentially the equilibrium one, given by $K_1[A][B]$, where K_1, the equilibrium constant for forming X from A and B, is equal to k_1/k_{-1}. The rate of reaction is then proportional to this equilibrium concentration of X and to that of C, so that the reaction will exhibit third order kinetics.

But when reaction step 2 is predominant, then virtually every molecule of the reaction intermediate X, once formed, is hastily scavenged by a molecule of C, so that the equilibrium concentration of X is never approached. In effect, step 1 has now become rate controlling, as evidenced by the form of equation (3.22a). Increasing the concentration of C cannot cause P to be formed any more quickly. It merely diminishes the steady state concentration of X, which is given by $k_1[A][B]/(k_2[C])$ and so is inversely proportional to the concentration of C.

3.4 The "third body" effect in atom recombination

When two atoms collide in the gas phase and attempt to combine chemically, the molecule they form must inevitably be one possessing a fatal amount of vibrational energy, as illustrated in Figure 3.3. The molecule cannot survive longer than the duration of one stretching vibration unless within this period of time it has been allowed to shed some of this excessive vibrational energy. This might be accomplished by a collision with another species, M. For the combination of two gaseous halogen atoms, the process could be written:

$$X + X + M \rightarrow X_2 + M \quad (3.23)$$

If no other substance has been added, the role of M could be played by an X atom or by X_2. If a large amount of another substance has been added, such contributions become swamped by that of the added substance.

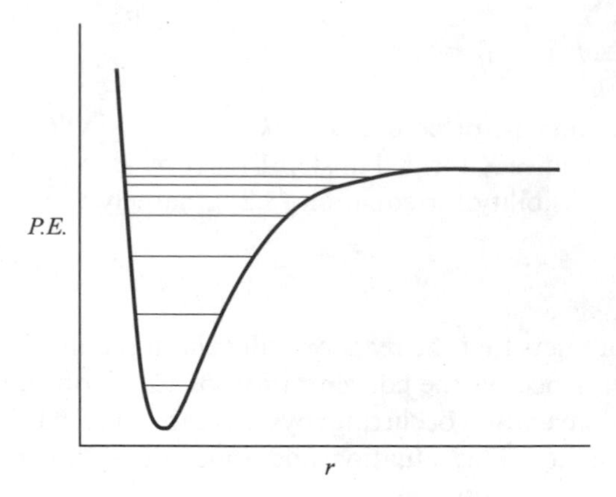

Fig. 3.3 Sketch of the vibrational levels (v) of a diatomic molecule in relation to the variation of the potential energy as a function of inter-nuclear separation, r.

Reaction (3.23) means that the process involves three species and implies that the formation of X_2 molecules should be kinetically second order with respect to X atoms and first order with respect to M, the third body. Since the concentration of the latter will remain constant, the reaction will thus follow second order kinetics, with the pseudo-second order rate constant proportional to the concentration of M.

From a practical point of view, M needs to make a collision during the *ca* 10^{-13} s of the first vibrational period of the fledgling X_2 molecule, and as a consequence to remove some of its vibrational energy. Efficiency in making the collision does not vary much with the size of the molecule since small molecules tend to have greater velocities. But in regard to energy transfer, whereas a helium atom can assist recombination only by acquiring extra translational energy, a polyatomic molecule may in addition gain rotational or vibrational energy, so a collision with this latter type will tend to be more effective.

Kinetic studies of atom recombination reactions show that these behave in accordance with reaction (3.23). Some typical values for the efficiency of different molecules in acting as the third body or *dreierstoss* are shown in Table 3.1, where the data relate to I atom recombination.

Two features of such reactions are worthy of special note. Firstly, the variations in the efficiencies of different molecules in acting as third bodies are considerable, and thus difficult to rationalise on the basis of variations in their size or in their readiness to accept vibrational energy. Secondly, while the recombination of I atoms in the presence of a third body usually proceeds slightly less quickly at higher temperatures, with the most efficient third bodies, reaction (3.23) has a highly negative activation energy.

An interpretation of this problem was offered by Porter, who proposed that the role of the third body may in some cases be a much more active one

Table 3.1 Rate constants and Arrhenius activation energies for various molecules acting as the third body in the reaction, $I + I + M \rightarrow I_2 + M$

M	$k/10^{-32}\,cm^6\,molec^{-2}\,s^{-1}$	$E_a/kJ\,mol^{-1}$
Helium	0.84	−5.9
Argon	1.64	−8.4
Oxygen	3.72	−9.2
Carbon dioxide	7.41	−10.0
Benzene	43.9	−10.0
Toluene	107	−14.2
Mesitylene	223	−20.1

Data taken from Porter, G. and J. A. Smith, 1961, *Proceedings of the Royal Society, A*, **261**, 28.

than is implied by reaction (3.23). Where it is possible for an I atom to form a weakly bound complex with the third body, then a mechanism analogous to (3.19) is possible, namely:

$$\left.\begin{array}{lcl} I + M & \overset{1}{\rightleftharpoons} & I.M \\ I.M + I & \overset{2}{\rightarrow} & I_2 + M \end{array}\right\} \qquad (3.24)$$

Spectroscopic evidence of such complexes in the gas phase has since been found, where M is benzene or a methyl-substituted benzene.

On this basis, the concentration of the weak complex is given by the equation

$$[I.M] = K_1[I][M] \qquad (3.25)$$

where K_1 denotes the equilibrium constant for step 1. Thus, for the rate of the reaction, we have,

$$\frac{d[I_2]}{dt} = k_2[I][I.M] = K_1 k_2 [I]^2[M] \qquad (3.26)$$

which is consistent with experiment.

Following equation (1.46), the Arrhenius activation energy of reaction (3.26) will be given by:

$$\begin{aligned} E_a &= -R\,\frac{d\ln(K_1 k_2)}{d(1/T)} \\ &= -R\,\frac{d\ln K_1}{d(1/T)} - R\,\frac{d\ln k_2}{d(1/T)} \qquad (3.27) \\ &= \Delta H_1^{\ominus} + E_2 \end{aligned}$$

where $\Delta H_1^{\ominus}$ denotes the standard enthalpy change for the formation of the weak complex, I.M, and E_2 is the activation energy of step 2. While the latter must be expected to be positive, $\Delta H_1^{\ominus}$ should be negative since the complex is bound, albeit weakly. Where $\Delta H_1^{\ominus}$ is more negative than E_2 is positive, the consequence will be a negative Arrhenius activation energy for the formation of I_2.

3.5 Parallel reactions

Let us consider the case where a reactant may engage in either of two reactions. For simplicity, we shall initially assume that both are unimolecular processes:

$$\left.\begin{array}{l} A \xrightarrow{1} B \\ A \xrightarrow{2} C \end{array}\right\} \tag{3.28}$$

The obvious consequence is that A will be used up in exactly the same way as it would be if it participated in a single reaction whose rate constant were the sum of k_1 and k_2,

$$-\frac{\mathrm{d}[A]}{\mathrm{d}t} = k_1[A] + k_2[A] = (k_1 + k_2)[A] \tag{3.29}$$

so that if the concentration of A is initially a and falls to $(a - x)$ at time t, then the integrated rate equation is given by:

$$\ln\left(\frac{a}{a-x}\right) = (k_1 + k_2)t \tag{3.30}$$

It is instructive to consider the rate of formation of one of the reaction products. If the concentration of B is initially zero, at time t let it be y and at $t = \infty$, let it be b. So for the rate of formation of B we have:

$$\frac{\mathrm{d}y}{\mathrm{d}t} = k_1(a - x) \tag{3.31}$$

However, b must be equal to a multiplied by $k_1/(k_1 + k_2)$, since this is the fraction of molecules A that will react by route 1 to yield B whereas the fraction $k_2/(k_1 + k_2)$ will react by route 2 to yield C. Likewise, $y = k_1x/(k_1+k_2)$. Thus we have:

$$\frac{\mathrm{d}y}{\mathrm{d}t} = k_1\left\{\left(\frac{k_1+k_2}{k_1}\right)b - \left(\frac{k_1+k_2}{k_1}\right)y\right\}$$

$$= (k_1 + k_2)(b - y) \tag{3.31a}$$

This equation, and its integrated form,

$$\ln\left(\frac{b}{b-y}\right) = (k_1 + k_2)t \tag{3.32}$$

show that if the build-up of one of the reaction products is observed, the rate constant deduced therefrom will not pertain to that route only but rather will represent the total rate constant for the two reactions in which the reactant may participate.

Thus any kinetic experiment performed on this system will yield only $(k_1 + k_2)$. The ratio of these rate constants may be found from the yields of the respective products,

$$\frac{k_1}{k_2} = \frac{\text{Yield of B}}{\text{Yield of C}} \tag{3.33}$$

and in this way the individual values of k_1 and k_2 may both be determined.

An example of a compound which, as in mechanism (3.28), may react by either of two routes, is isopropenylcyclobutane, which by one route yields 1-methylcyclohexene and by the other, equimolar quantities of ethene and isoprene. A solution phase example is the hydrolysis in water at around 100°C of 1-chloropropan-2-ol, which yields HCl along with either acetone or propan-1,2-diol.

Alternatively, we may consider two competing reactions for a substrate, of which one is unimolecular and the other is a bimolecular one involving another reagent.

$$\left.\begin{array}{lcl} \text{A} & \xrightarrow{1} & \text{C} \\ \text{A} + \text{B} & \xrightarrow{2} & \text{D} \end{array}\right\} \tag{3.34}$$

The rate of reaction of A is then given by

$$\begin{aligned} -\frac{\text{d[A]}}{\text{d}t} &= k_1[\text{A}] + k_2[\text{A}][\text{B}] \\ &= (k_1 + k_2[\text{B}])[\text{A}] \end{aligned} \tag{3.35}$$

If B is present in large excess, reaction 2 will follow pseudo-first order kinetics, with the pseudo-first order rate constant equal to k_2[B]. In this case the parallel reactions of (3.34) will be kinetically similar to those of (3.28), but k_1 and k_2 may be determined individually from the manner in which the pseudo-first order rate constant, equal to $(k_1 + k_2[\text{B}])$, varies with the value of the excess concentration of B.

A reaction to which this may be applicable is the hydrolysis of an alkyl halide, RX, in alkaline solution. In some instances both S_N1 and S_N2 mechanisms may be operative under these conditions and in order to determine their respective contributions it is necessary to measure the pseudo-first order rate constant, k', of the hydrolysis of RX as a function of the excess concentration of hydroxide ion. When k' is then plotted against OH^- concentration, the intercept at zero OH^- concentration is the rate constant for the unimolecular S_N1 reaction and the slope is the rate constant of the bimolecular S_N2 process.

Another possible scenario is that the reactant may react either by a unimolecular process or by a bimolecular reaction with itself.

$$\left.\begin{array}{lcl} \mathrm{A} & \xrightarrow{1} & \mathrm{B} \\ \mathrm{A+A} & \xrightarrow{2} & \mathrm{C} \end{array}\right\} \tag{3.36}$$

Thus the kinetic equation is:

$$\frac{\mathrm{d}x}{\mathrm{d}t} = k_1(a - x) + 2k_2(a - x)^2 \tag{3.37}$$

The integration of this equation yields:

$$k_1 t = \ln\left(\frac{a[k_1 + 2k_2(a - x)]}{(a - x)(k_1 + 2k_2 a)}\right) \tag{3.38}$$

Rearranging this equation, we obtain:

$$x = \frac{a(k_1 + 2k_2 a)(\mathrm{e}^{k_1 t} - 1)}{k_1 \mathrm{e}^{k_1 t} + 2k_2 a(\mathrm{e}^{k_1 t} - 1)} \tag{3.39}$$

While this is an explicit expression for x in terms of a and the two rate constants, the form of the equation is such that, knowing the manner in which the concentration of A decreases with time, it is not a simple matter to determine the values of the individual rate constants, k_1 and $2k_2$.

On the basis of equation (3.39), calculated values of the concentration, as a function of time, of a species which undergoes both first order and second order reactions, as in mechanism (3.36), are shown in Figure 3.4. Figure 3.5 shows the relevant first order and second order plots, using these same data. It may readily be seen that, while neither is linear, the second order plot is reasonably so over the first part of the decay and likewise the first order plot over the last part. This illustrates that the bimolecular reaction is relatively more important at high concentrations and the unimolecular reaction at low concentrations.

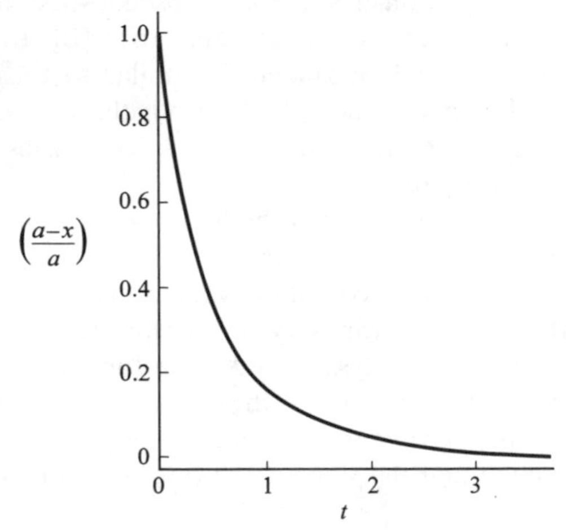

Fig. 3.4 Plot of reactant concentration against time based on equation (3.39), referring to two reactions of respective orders one and two.

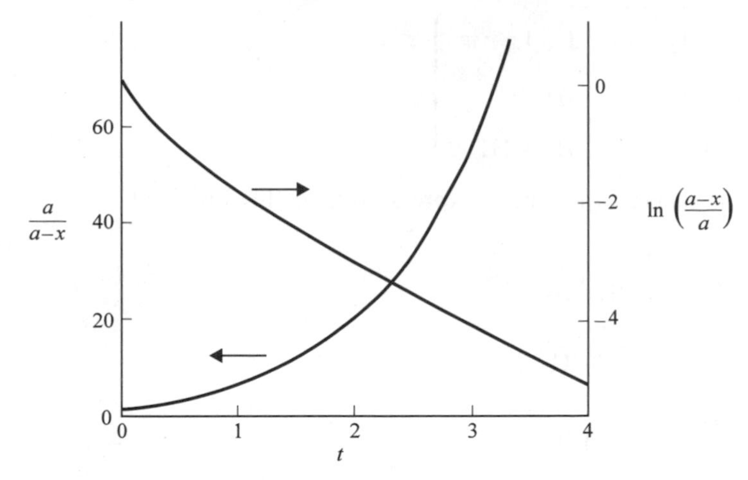

Fig. 3.5 First and second order plots based on the data of Figure 3.4. Left hand scale, $a/(a-x)$; right hand scale, $\ln\{(a-x)/a\}$.

3.6 Reactant participating in equilibria

In some chemical reactions the species which participates in the initial step of the reaction is not the normal form of the reactant, but one which is obtained from it by a dissociative or associative equilibrium. Such equilibria may cause the reaction order to appear a little unusual. However in some cases other steps in the mechanism may have the effect of obscuring these unusual aspects.

The gas phase reaction of I_2 with H_2, which has a very simple stoichiometric equation,

$$H_2 + I_2 = 2HI \tag{3.40}$$

was studied by Bodenstein in 1899 and found to be kinetically first order with respect to each reactant. On this basis it was generally assumed to be an elementary bimolecular process and some textbooks featured diagrams of the four-centre transition state that was presumed to be involved in this reaction.

As a consequence of photochemical studies by Sullivan in 1967, it was shown that the reaction rate that would be achieved via a non-elementary route could fully account for the kinetics of the reaction and for the reaction rate observed. This showed that what had been generally assumed in regard to the reaction mechanism was of (3.40) in fact incorrect.

Among the halogens, iodine has the weakest X–X bond, so that the dissociation of I_2 into atoms occurs quite readily. An iodine atom can react with H_2 to form the radical, IH_2, and when this complex meets an I atom, one possible outcome is the formation of two molecules of HI. Thus a probable mechanism is:

$$\left.\begin{array}{rcl} I_2 & \overset{1}{\rightleftharpoons} & I + I \\ I + H_2 & \overset{2}{\rightleftharpoons} & IH_2 \\ I + IH_2 & \overset{3}{\rightarrow} & HI + HI \end{array}\right\} \tag{3.41}$$

From this scheme, the kinetic expression is readily obtained:

$$[I] = K_1^{1/2}[I_2]^{1/2} \tag{3.42}$$

$$[IH]_2] = K_2[H_2][I]$$

$$= K_1^{1/2}K_2[H_2][I_2]^{1/2} \tag{3.43}$$

$$\frac{d[HI]}{dt} = 2k_3[IH_2][I]$$

$$= 2k_3K_1K_2[H_2][I_2] \tag{3.44}$$

So on the basis of reaction scheme (3.41), the reaction should follow second order kinetics.

Another example is afforded by the reactions in solution of a species A which undergoes association. Supposing there are n molecules in the aggregate so formed,

$$nA \rightleftharpoons A_n \tag{3.45}$$

then, if the equilibrium constant is K_n, we have:

$$[A] = K_n^{-1/n}[A_n]^{1/n} \tag{3.46}$$

Suppose equilibrium (3.46) lies well towards the right hand side, so that almost exclusively, the reactant A is present as part of an aggregate. The concentration of A_n is then, to a good approximation, $1/n$ times the nominal concentration of A, given by the number of moles of A present divided by the volume of the solution. If we call this nominal concentration c, then that of A_n is given by c/n, and so the actual concentration of A is given by:

$$[A] = K_n^{-1/n}\left(\frac{c}{n}\right)^{1/n} \tag{3.46a}$$

Butyl lithium acts as an initiator of anionic polymerisation, with the rate showing an apparent dependence on the initiator concentration to the power 1/6. The accepted explanation is that in these solutions butyl lithium is associated, forming a hexamer, so that the actual concentration of the initiator is proportional to its nominal concentration to the power one-sixth.

Another possibility is that a reactant species may participate in acid–base equilibria,

$$A \rightleftharpoons B^- + H^+ \tag{3.47}$$

Where spectrophotometry is used to follow the reaction, it will typically enable the concentration of either A or B^- to be measured. If the absorption peak being monitored were due to A, but the species taking part in the reaction were B^-, then it is relevant that the actual concentration of B^- is given by,

$$[B^-] = K_a[A]/[H^+] \tag{3.48}$$

where K_a is the dissociation constant for reaction (3.47).

On the other hand, if the reaction were slow, it might be appropriate to measure reactant concentration by titrimetric analysis on aliquots, which would normally measure the combined concentrations of A and B^-. If this is denoted by a, then the actual concentration of B^- is given by

$$[B^-] = \frac{aK_a}{K_a + [H^+]} \tag{3.49}$$

and that of A is given by

$$[A] = \frac{a[H^+]}{K_a + [H^+]} \tag{3.50}$$

If both A and B^- react, but at quite different rates, then there will be distinctive variations in the rate with the H^+ concentration. For simplicity, let us assume the reactions are unimolecular:

$$\left.\begin{array}{lllll} A & \xrightarrow{1} & P & & \\ B^- & \xrightarrow{2} & Q & \xrightarrow{H^+} & P \end{array}\right\} \tag{3.51}$$

This leads to the equation:

$$\begin{aligned} \frac{d[P]}{dt} &= k_1[A] + k_2[B^-] \\ &= \frac{\{k_1[H^+] + k_2K_a\}a}{K_a + [H^+]} \\ &= k'a \end{aligned} \tag{3.52}$$

At a constant concentration of H^+, this equation describes simple first order kinetics. The variation of the apparent first order rate constant, k', with the pH is illustrated in Figure 3.6.

3.7 Opposing reactions

The simplest reaction scheme involving opposing reactions would relate to a unimolecular isomerisation reaction. By the Principle of Microscopic Reversibility, the reverse reaction would also be a unimolecular process:

$$\left.\begin{array}{lll} A & \xrightarrow{1} & B \\ B & \xrightarrow{-1} & A \end{array}\right\} \tag{3.53}$$

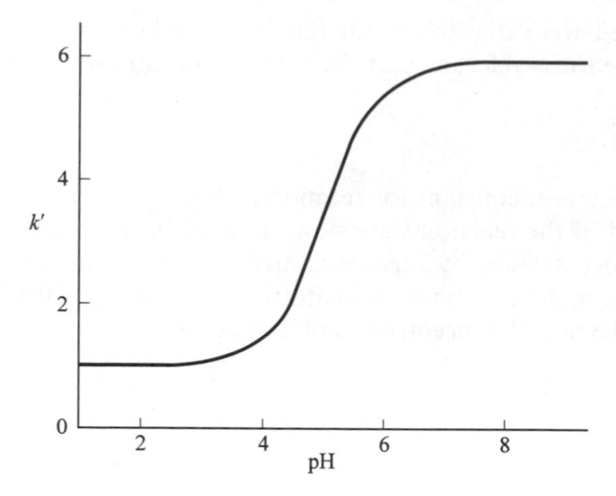

Fig. 3.6 Variation of the apparent first order rate constant, k', with the pH, calculated from equation (3.52) putting $k_1 = 1$, $k_2 = 6$ and $K_a = 10^{-5}$.

The essence of this principle is that where a system is in dynamic equilibrium, the forward and reverse rates of each individual step must be equal.

Starting with a concentration of A equal to a and of B equal to zero, the kinetic equation is:

$$\frac{dx}{dt} = k_1(a - x) - k_{-1}x \tag{3.54}$$

This process, unlike all of those considered previously, will not proceed to the point where the reactant A is totally used up, but rather it advances towards an equilibrium concentration of A. At that stage, dx/dt will be zero and it is convenient to denote this final value of x as x_e. At that point, we have:

$$k_1(a - x_e) = k_{-1}x_e \tag{3.55}$$

Substituting into equation (3.54), we obtain:

$$\frac{dx}{dt} = (k_1 + k_{-1})(x_e - x) \tag{3.56}$$

Integrating and putting $x = 0$ at $t = 0$, we have the equation

$$\ln\left(\frac{x_e}{x_e - x}\right) = (k_1 + k_{-1})t \tag{3.57}$$

Thus a reaction based on scheme (3.53) will follow an equation closely resembling that for a simple first order reaction considered in Chapter 1. The more obvious difference between equation (1.15) and (3.57) is that the concentration, a, of A in the former is replaced in the latter by x_e, which represents the maximum extent to which this concentration will now be depleted. The other change is that the factor in front of t is now the sum of the forward and the reverse rate constants. A kinetic study will thus yield $(k_1 + k_{-1})$. Knowledge of the position of equilibrium can be used to find k_1/k_{-1} and thus to permit both to be evaluated.

Many isomerisation reactions are kinetically first order or pseudo-first order processes, so that equation (3.57) is applicable to their approach to equilibrium. This is illustrated in Example 3.1.

Example 3.1

The following data refer to the reversible isomerisation of stilbene, which exists in *cis* and *trans* forms:

Time/min	0	20	50	80	120	170	∞
%age *cis*	100	92.5	82.3	73.6	63.7	53.8	17.1

Evaluate the rate constants for the processes, *cis* → *trans* and *trans* → *cis* under these conditions.

Answer. Since equation (3.55) involves only *ratios* of concentrations, it is convenient to use percentages of the total amount of stilbene as a measure of the concentration. Thus $x_e = 82.9$, and from the data given one may calculate as follows:

Time/min	0	20	50	80	120	170	∞
$(x_e - x)$	82.9	75.4	65.4	56.5	46.6	36.7	0
$\ln(x_e - x)$	4.42	4.32	4.18	4.03	3.84	3.60	

The plot of $\ln(x_e - x)$ against time is shown in Figure 3.7.

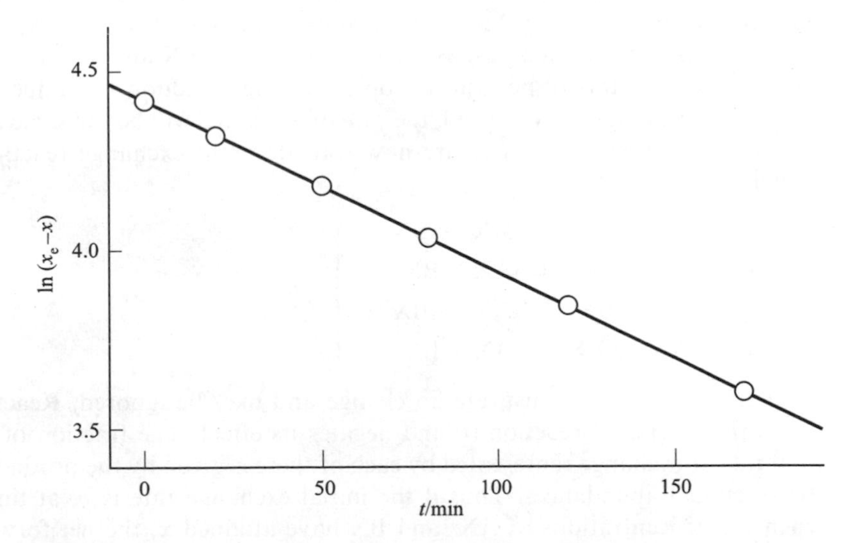

Fig. 3.7 Plot of $\ln(x_e - x)$ against time, using the data of Example 3.1.

From the slope of this graph, $(k_1 + k_{-1})$ may be evaluated as $4.8 \times 10^{-3}\ \text{min}^{-1}$. At equilibrium, $k_1 \times 17.1 = k_{-1} \times 82.9$. Thus $k_1 = 4.85k_{-1}$. This means that $5.85k_{-1} = 4.8 \times 10^{-3}\ \text{min}^{-1}$, so we have:

$$k_{-1} = 0.82 \times 10^{-3}\ \text{min}^{-1}$$
$$k_1 = 3.98 \times 10^{-3}\ \text{min}^{-1}$$

3.8 Isotopic exchange reactions

It is convenient to include isotopic exchange reactions in this chapter, even though the reaction processes involved need not be a simple combination of elementary steps.

Suppose X^* denotes a distinctive isotope of the element X, that A and B are atoms or radicals each of which combines with X to form a molecule and that exchange of the X atom may occur between AX and BX. Let us consider the reaction:

$$AX + BX^* \rightleftharpoons AX^* + BX \tag{3.58}$$

If we make the simplifying assumption (which is not in general strictly true) that the forward and reverse reactions occur with the same rate constant, then we may calculate the equilibrium concentrations of all four species on the basis that the equilibrium constant is 1. If the initial concentration of AX is a, and that of BX^* is b, with no AX^* or BX present initially, then the equilibrium concentrations are respectively $a^2/(a + b)$, $b^2/(a + b)$ and, for each of the products, $ab/(a + b)$.

The manner in which these equilibrium concentrations are approached is a consequence of the fact that, as the reaction proceeds, the concentrations that change are only those of isotopic make-up. The sum of those of AX and AX^* remains constant at a, just as the sum of those of BX and BX^* remains constant at b. In this respect, an isotopic exchange reaction is unique. So, after some exchange has occurred, the rate of exchange will be the same as it was initially, except that there are now four different exchange reactions, namely:

$$\left.\begin{aligned}
&\text{(i)} & AX + BX^* &= AX^* + BX \\
&\text{(ii)} & AX + BX &= AX + BX \\
&\text{(iii)} & AX^* + BX^* &= AX^* + BX^* \\
&\text{(iv)} & AX^* + BX &= AX + BX^*
\end{aligned}\right\} \tag{3.59}$$

Of these, (ii) and (iii) constitute no change and may be ignored. Reaction (iv) is the reverse of reaction (i) and negates its effect. The fraction of the total rate of exchange represented by each of these is given by the product of the fractional abundances. Thus if the initial exchange rate is ρ, at time t when the concentrations of AX^* and BX have attained x, the net forward rate of exchange may be expressed as follows:

$$\frac{dx}{dt} = \rho\left\{\left(\frac{a-x}{a}\right)\left(\frac{b-x}{b}\right) - \frac{x^2}{ab}\right\}$$

$$= \rho\left(\frac{a+b}{ab}\right)\left(\frac{ab}{a+b} - x\right)$$

$$= \rho\left(\frac{a+b}{ab}\right)(x_e - x) \tag{3.60}$$

Integrating and using the fact that $x = 0$ when $t = 0$, we obtain the equation:

$$\ln\left(\frac{x_e}{x_e - x}\right) = \rho\left(\frac{a+b}{ab}\right)t \tag{3.61}$$

This equation resembles equations (3.57) and (1.15), which were both derived for reactions which were assumed to be first order processes. In this case, we have made no assumption in regard to the actual reaction order. The form of equation (3.61) arises not from the kinetic order of the reaction involved, but rather from the fact that, as a reaction like (3.58) takes place, it is only the isotopic composition and not the concentration which changes.

While the manner in which equilibrium is approached in one experiment gives no indication of the kinetic order, it is possible to determine this from a series of experiments with differing initial concentrations. Let us suppose that reaction (3.58) is of order p with respect to AX and order q with respect to BX^*. Thus ρ may be equated to

$$\rho = k_n a^p b^q \tag{3.62}$$

so that the integrated rate equation becomes:

$$\ln\left(\frac{x_e}{x_e - x}\right) = k_n(a+b)a^{p-1}b^{q-1}t \tag{3.62a}$$

Thus when $\ln(x_e - x)$ is plotted against t, the pseudo-first order rate constant so obtained, k', may be equated to $k_n(a+b)a^{p-1}b^{q-1}$. Rearranging and taking logarithms we have

$$\ln\frac{k'}{(a+b)} = \ln k_n + (p-1)\ln a + (q-1)\ln b \tag{3.63}$$

So if the concentration of AX is varied while that of BX^* is kept constant, the plot of $\ln[k'/(a+b)]$ against $\ln a$ enables $(p-1)$ to be determined from the slope. Similarly, q may be found if the concentration of BX^* is varied with that of AX kept constant.

In many cases, the molecule undergoing isotopic substitution has more than one X atom. Suppose the species undergoing exchange with BX^* is AX_3, where all three X atoms are equivalent. We may now have three separate products, AX_2X^*, AXX^*_2 and AX^*_3. If these species have attained the respective concentrations u, v and w, then the extent of isotopic exchange may be represented by ϕ, defined as:

$$\phi = u + 2w + 3w \tag{3.64}$$

It can readily be shown that the integrated equation for the reaction of BX^* with AX_3 is

$$\ln\left(\frac{\phi_e}{\phi_e - \phi}\right) = \frac{\rho}{\phi_e} t \tag{3.65}$$

which is totally analogous to equation (3.61).

Earlier, it was assumed that the equilibrium constant for reaction (3.58) is one. Actually, the effect of isotopic substitution on such a system is not in general negligible, reflecting the fact that a reaction involving one isotope may occur at a different rate from that involving another (see Section 5.6). Consequently, the actual value of K needs to be known before x_e can be calculated. However, even with an isotope effect, the manner in which x approaches x_e is such that $\ln(x_e - x)$ is essentially a linear function of t, so that equation (3.61) remains applicable, and likewise equation (3.63) for multiple exchange.

Example 3.2

The three hydrazyl hydrogen atoms of methyl hydrazine undergo exchange with deuterium over a palladium catalyst at 273 K, while the methyl hydrogen atoms do not. In an experiment with 10 torr of $MeNHNH_2$ and 40 torr of D_2, the amounts of the deuterated products, $MeN_2H_2D(=d_1)$, $MeN_2D_2(=d_2)$ and $MeN_2D_3(=d_3)$ increased with time as follows:

t/min	0	5	10	15	20	25	30
d_0	67.1	64.0	61.2	58.6	56.4	54.0	51.8
d_1	10.1	13.0	15.3	17.4	19.1	20.9	22.3
d_2	0.5	0.8	1.2	1.7	2.2	2.8	3.4
d_3	–	–	–	–	0.1	0.1	0.2

Evaluate the pseudo-first order rate constant for the deuterium exchange reaction under these conditions.

Answer. It is convenient to work with the ϕ values as fractions, in the range $0 < \phi < 1$. On that basis,

$$\phi_e = \frac{40 \times 2}{(40 \times 2) + (10 \times 3)} = 0.727$$

That is, since the D atoms present constitute 0.727 of the atoms in the "dihydrogen" plus the hydrazyl "hydrogens", assuming no isotope effect we conclude that this is the fraction of the hydrazyl "hydrogens" which, at equilibrium, would be D atoms.

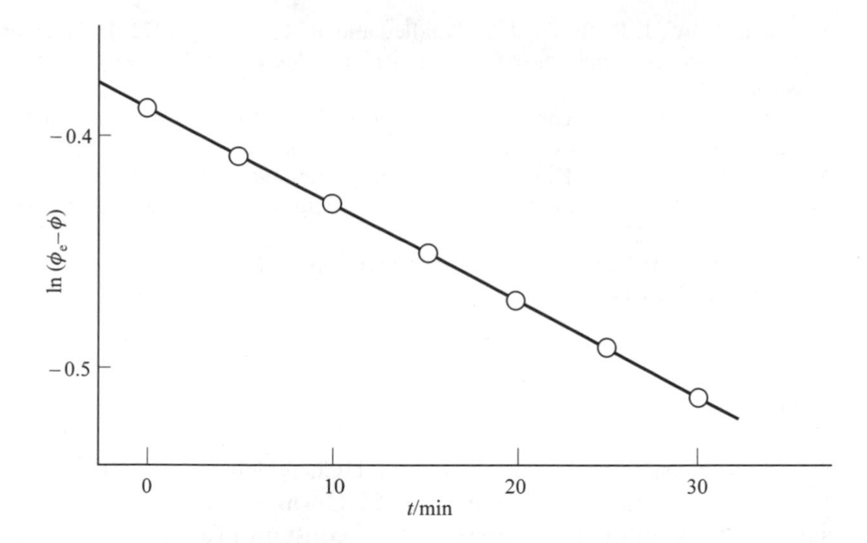

Fig. 3.8 Plot of $\ln(\phi_e - \phi)$ against time, using data of Example 3.2.

Now we evaluate $\phi = (d_1 + 2d_2 + 3d_3)/3(d_o + d_1 + d_2 + d_3)$

t/min	0	5	10	15	20	25	30
ϕ	0.048	0.062	0.076	0.089	0.102	0.115	0.127
$(\phi_e - \phi)$	0.679	0.665	0.651	0.638	0.625	0.612	0.600
$\ln(\phi_e - \phi)$	−0.387	−0.408	−0.429	−0.449	−0.470	−0.491	−0.511

The plot of $\ln(\phi_e - \phi)$ against t is shown in Figure 3.8. From the slope, the rate constant for deuterium exchange is found to be 4.16×10^{-3} min^{-1}.

Suggested reading

Viossat, V. and R. I. Ben-Aim, 1993. A test of the validity of steady state and equilibrium approximations in chemistry. *Journal of Chemical Education*, **70**, 732–8.

McCracken, D. R. and G. V. Buxton, 1981. Failure of Arrhenius equation for hydroxyl radical-bicarbonate ion reaction above 100°C. *Nature*, **292**, 439–41.

Tardy, D. C. and E. D. Cater, 1983. The steady state and equilibrium assumptions in chemical kinetics. *Journal of Chemical Education*, **60**, 109–11.

Porter, G. and J. A. Smith, 1961. The recombination of atoms. III Temperature coefficients for iodine atom recombination. *Proceedings of the Royal Society*, *A*, **261**, 28–37.

Sullivan, J. H., 1967. Mechanism of the "bimolecular" hydrogen-iodine reaction. *Journal of Chemical Physics*, **46**, 73–8.

Brosseau, V. A., J. R. Basila, J. F. Smalley and R. L. Strong, 1972. Halogen atom charge-transfer complexes in the vapour phase. *Journal of the American Chemical Society*, **94**, 716–19.

Edwards, J. O., E. F. Greene and J. Ross, 1968. From stoichiometry and rate law to mechanism. *Journal of Chemical Education*, **45**, 381–5.

McKay, H. A. C., 1938. Kinetics of exchange reactions. *Nature*, **142**, 997–8.

Logan, S. R., 1990. The kinetics of isotopic exchange reactions. *Journal of Chemical Education*, **67**, 371–3.

Espenson, J. H., 1995. *Chemical Kinetics and Reaction Mechanisms*, 2nd edition. New York: McGraw-Hill.

Problems

3.1 In the reaction between NO_2 and HCl, which yields NOCl, Cl_2 and H_2O, the initial rate of reaction (ρ) at 55°C was measured at various pressures of NO_2 with the HCl pressure kept constant (Talbot, P. J. and J. H. Thomas, *Trans Faraday Soc.*, 1959, **55**, 1884–91).

P_{NO_2}/torr	7.4	7.5	8.6	12.3	14.8	17.5	24.3
ρ/torr min^{-1}	1.12	1.48	1.62	3.2	4.4	6.6	13.8

Hence determine the reaction order with respect to NO_2. Bearing in mind the dimer–monomer equilibrium,

$$N_2O_4 \rightleftharpoons 2NO_2$$

with the consequence that ${P_{NO_2}}^2 = K_p P_{N_2O_4}$, express the dependence of the initial rate on $P_{N_2O_4}$.

3.2 As shown in Figures 3.1 and 3.2, the concentration profile of the intermediate product B has a point of inflexion at a time t_f. Starting with equation (3.10), find an expression for t_f in terms of k_1 and k_2 . How is t_f related to t_m, the point where the maximum concentration of B is achieved?

3.3 Aromatic peroxyacids, RCO_3H, slowly decompose in aqueous solution to the aromatic carboxylic acid and oxygen. The rate is strongly pH-dependent.

$$RCO_3H \;=\; RCO_2H \;+\; \tfrac{1}{2}O_2$$

In kinetic studies with peroxybenzoic acid, $PhCO_3H$ ($pK_a = 7.78$) (Goodman, J. F., P. Robson and E. R. Wilson, 1962, *Trans Faraday Soc.*, **58**, 1846–51), the concentration was determined by iodometric analysis of aliquots. Plots of the reciprocals of these concentrations against time gave good straight lines. The value of the slope, termed k_o, was found to vary with the pH of the solution as follows overpage. From these data, deduce the identity of the reaction process, evaluate its rate constant and propose a detailed reaction mechanism.

pH	$k_0/10^3\ dm^3\ mol^{-1}\ s^{-1}$	pH	$k_0/10^3\ dm^3\ mol^{-1}\ s^{-1}$
6.6	1.2	7.9	4.5
6.95	1.6	8.1	4.6
7.07	2.2	8.16	3.5
7.18	3.2	8.41	2.8
7.45	3.6	8.55	2.1
7.58	4.4	8.8	1.8
7.7	4.5	9.0	1.2

3.4 Photo-excitation of benzophenone in aqueous 2-propanol leads to the formation of the ketyl radical, $Ph_2\dot{C}OH$, which deprotonates to give the ketyl radical anion,

$$Ph_2\dot{C}OH \rightleftharpoons Ph_2CO^- + H^+$$

with the pK_a value equal to 9.2.

In flash photolysis experiments with the pH in excess of 10, the transient absorption at 630 nm due to the ketyl radical anion was found to decay by second order kinetics with the slope of the plot of $[PhCO^-]^{-1}$ against time giving the following values for the rate constant:

pH	$k'/dm^3\ mol^{-1}\ s^{-1}$
10.1	1.43×10^8
11.3	8.25×10^6
11.9	1.56×10^6
12.9	2.3×10^5

Interpret these results in terms of the possible radical reactions,

$$Ph_2\dot{C}OH + Ph_2\dot{C}OH \rightarrow \text{dimer}$$

$$Ph_2\dot{C}OH + Ph_2CO^- \rightarrow \text{dimer}$$

$$Ph_2CO^- + Ph_2CO^- \rightarrow \text{dimer}$$

reaching whatever conclusions are justified. (Beckett, A. and G. Porter, *Trans Faraday Soc*., 1963, **59**, 2038–50.)

3.5 In the catalytic reaction between ethylamine (5 torr) and D_2 (30 torr) under conditions where only the amine hydrogen atoms underwent exchange, the amounts of the various ethylamines, $EtNH_2(=d_0)$, $EtNHD(=d_1)$, and $EtND_2(=d_2)$ were found to vary with time as follows:

t/min	0	5	10	15	20	25
d_0	68.4	63.5	58.7	54.3	50.3	46.6
d_1	3.2	8.1	12.4	16.3	19.3	22.2
d_2	–	0.2	0.6	1.1	1.9	2.7

Evaluate the pseudo-first order rate constant for the deuterium exchange reaction under these conditions.

3.6 The compounds A and B are isomers which undergo inter-conversion. Starting with a sample of pure A, the concentration was found to decrease with time as follows:

t/min	0	1.5	3.0	5.0	8.0	11.0	∞
$[\text{A}]/\text{mol dm}^{-3}$	0.100	0.080	0.065	0.050	0.037	0.029	0.018

Thus deduce the rate constants for both isomerisation processes, assuming that both reactions are first order or pseudo-first order.

3.7 There are two rival mechanisms for the formation of phosgene, $COCl_2$, from CO and Cl_2:

I

$$Cl_2 \rightleftharpoons 2Cl \qquad K_1$$

$$Cl + CO \rightleftharpoons COCl \qquad K_2$$

$$COCl + Cl_2 \xrightarrow{3} COCl_2 + Cl$$

$$COCl_2 + Cl \xrightarrow{4} COCl + Cl_2$$

II

$$Cl_2 \rightleftharpoons 2Cl \qquad K_1$$

$$Cl + Cl_2 \rightleftharpoons Cl_3 \qquad K_5$$

$$Cl_3\cdot + CO \xrightarrow{6} COCl_2 + Cl$$

$$Cl\cdot + COCl_2 \xrightarrow{7} CO + Cl_3$$

For each mechanism, derive an expression for the rate of formation of phosgene in terms of the concentrations of CO and Cl_2 and relevant rate constants and show how one might decide which mechanism is correct.

3.8 Strontium-90 ($^{90}_{38}$Sr) undergoes β^- emission with a half-life of 28 years to yield yttrium-90 ($^{90}_{39}$Y), which is also radioactive and has a half-life of 64 hours. If a sample of 100 g of pure ^{90}Sr is placed in a container, how much ^{90}Y will be present (a) after 24 hours and (b) after 24 years?

4 Theories of bimolecular reactions

As was indicated in the previous chapter, the majority of individual reaction steps involve the participation of two reactant molecules. Thus a thorough understanding of bimolecular reactions is paramount in seeking to comprehend chemical kinetics.

In this chapter we look at the two fundamental theories of bimolecular reactions, the Collision Theory and the Transition State Theory, and compare their results. Then we look at the effect a solvent has on bimolecular collisions and consider the kinetic behaviour of diffusion controlled, and nearly diffusion-controlled, reactions in solution.

4.1 The Collision Theory

The origins of the Collision Theory, during the second decade of this century, owe something to the currency at that time of the rather controversial "Radiation Hypothesis", that chemical reaction comes about through the absorption of infra-red radiation. The study that W. C. McC. Lewis made, sufficed to show that reaction rates could easily be interpreted in terms of inter-molecular collisions and so the hypothesis became redundant.

4.1.1 The frequency of inter-molecular collisions

In attempting to derive an expression for the rate of the bimolecular reaction of A with B, it would be prudent initially to confine attention to the gas phase, in which the behaviour of inter-molecular collisions is well understood. The first objective is to evaluate the number of A–B collisions that take place in unit volume in unit time, assuming certain concentrations of A and of B.

For present purposes it is conventional to treat the molecules A and B as if they are smooth spheres, of respective radii r_A and r_B. An A–B collision thus involves the centres of the two molecules coming to a separation $(r_A + r_B)$, which we shall call d. Consider one molecule of A, moving with its mean velocity, $\bar{c}$, given, on the basis of the Kinetic Theory of Gases by the equation,

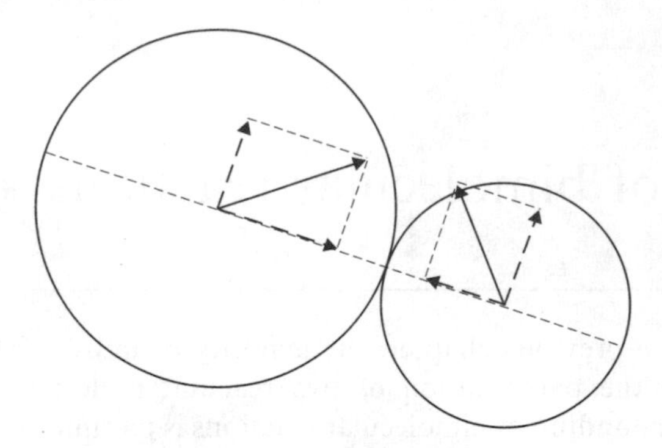

Fig. 4.1 Diagram showing the relevant components of velocity when molecules A and B undergo collision in the gas phase.

$$\bar{c} = \left(\frac{8k_B T}{\pi m_A}\right)^{1/2} \tag{4.1}$$

where k_B denotes the Boltzmann constant, T the temperature and m_A the mass of a molecule of A. In unit time this molecule, in moving the distance $\bar{c}$, will make collision with every B molecule whose centre lies within a distance d of the path of the centre of the A molecule, as sketched in Figure 4.1. The number of B molecules involved will be all those whose centres lie within the volume $\pi d^2 \bar{c}$.

However, while this answer would be correct if all the B molecules were stationary, the fact that they are all in motion means that an even greater number of collisions will take place. The improved answer is, to a good approximation, obtained by using, instead of $\bar{c}$, the mean velocity of a molecule of A relative to that of a molecule of B, $\bar{c}_{rel}$. This is obtained by using, in equation (4.1), the reduced mass $\mu = m_A m_B/(m_A + m_B)$ instead of the mass of an A molecule.

To obtain the number of A–B collisions made by this A molecule in unit time, we need to multiply the volume, $\pi d^2 \bar{c}_{rel}$, by n_A, the mean number of A molecules per unit volume. For the total number of A–B collisions per unit volume per unit time, N_{AB}, this product needs to be multiplied by n_B, the mean number of B molecules per unit volume, leading to the equation,

$$\begin{aligned} N_{AB} &= n_A n_B d^2 \pi \left(\frac{8(m_A + m_B)k_B T}{\pi m_A m_B}\right)^{1/2} \\ &= n_A n_B d^2 \left[\frac{8\pi(M_A + M_B)RT}{M_A M_B}\right]^{1/2} \end{aligned} \tag{4.2}$$

where R, the gas constant, and M_A, the relative molar mass of A, are obtained by multiplying above and below by Avogadro's constant, N_A.

The above deduction, using as it does the Collision Theory of gases, refers to collisions in the gas phase. In solution, the motion of the reactant molecules is constrained by the intervening solvent molecules, but most studies have come to the conclusion that the actual number of A–B collisions in solution is very similar to the number in the gas phase with the same reactant concentrations. However, as we shall see in a later section, there are some differences in the nature of the collisions.

It is instructive to digress at this point to consider the corresponding number of A–A collisions per unit time. This may be approached in the same manner as above, taking d_A as the diameter of an A molecule, but the product of $\pi d_A^2 \bar{c}_{rel} n_A$ and n_A counts every A–A collision twice and so needs to be divided by two. This gives:

$$N_{AA} = n_A^2 d_A^2 \left[\frac{2\pi(M_A + M_A)RT}{M_A^2}\right]^{1/2} \tag{4.3}$$

which shows that, if A and B were to have equal masses and radii, then with equal concentrations of A and B there would be twice as many A–B collisions as A–A collisions.

Other things being equal, the rate of reaction must be proportional to the number of relevant collisions per unit volume per unit time. A comparison of equations (4.2) and (4.3) shows that, for the rate constants of these reactions to be directly comparable, one should take $2k$ for the A–A reaction and k for the A–B reaction. In solution kinetics there is such a convention: it is, for example, normal to cite k for the reaction of OH + Br^-, but $2k$ for the reaction of OH + OH.

From the definitions of reaction rate and of rate constant, explained in Chapter 1, the rate equation for the bimolecular reaction of A + B is conventionally written,

$$-\frac{d[A]}{dt} = k[A][B] \tag{4.4}$$

whereas for the bimolecular reaction of A + A, the rate equation is:

$$-\frac{d[A]}{dt} = 2k[A]^2 \tag{4.5}$$

However, for the reasons advanced immediately above, the parameter $2k$ of equation (4.5) is directly comparable to the k of equation (4.4).

From equation (4.2) we may derive the collision number, Z_{AB}, which is equal to the second order rate constant that would pertain if every A–B collision were to lead to reaction. This is obtained by dividing N_{AB} by the concentrations of A and of B. The units of the reaction rate are then molecules per unit volume per unit time, so we have:

$$Z_{AB}/\text{m}^3\,\text{molec}^{-1}\,\text{s}^{-1} = d^2\left[\frac{8\pi(M_A + M_B)RT}{M_A M_B}\right]^{1/2} \tag{4.6}$$

where all quantities on the r.h.s. are in SI units. From this we may obtain the collision number in the units which are more conventional.

(i) in the gas phase,

$$Z_{AB}/\text{cm}^3\ \text{molec}^{-1}\ \text{s}^{-1} = 10^6 d^2 \left[\frac{8\pi(M_A + M_B)RT}{M_A M_B}\right]^{1/2} \tag{4.6a}$$

and (ii) in solution,

$$Z_{AB}/\text{dm}^3\ \text{mol}^{-1}\ \text{s}^{-1} = 10^3 N_A d^2 \left[\frac{8\pi(M_A + M_B)RT}{M_A M_B}\right]^{1/2} \tag{4.6b}$$

where N_A denotes Avogadro's constant.

We may use these equations to evaluate the collision number for $N_2 + O_2$ at 298 K, taking the respective molecular radii as 0.157 and 0.147 nm, which gives the collision diameter $d = 3.04 \times 10^{-10}$ m. This comes to $1.89 \times 10^{-10}\ \text{cm}^3\ \text{mol}^{-1}\ \text{s}^{-1}$ or to $1.14 \times 10^{11}\ \text{dm}^3\ \text{mol}^{-1}\ \text{s}^{-1}$.

4.1.2 The fraction of collisions in which reaction occurs

The fraction of A–B collisions in which a chemical reaction can occur is an issue which posed some problems in the 1920s. Clearly it is not simply a question of the Maxwell–Boltzmann distribution of velocities: there is no reason *a priori* why molecules with velocity greater than a certain value can react while those with a lower velocity may not.

One approach is to postulate that, for reaction to occur, there must at least be a certain amount of vigour to the A–B collision. Looking, in Figure 4.1, at the colliding pair of molecules, we see that since they are considered as spheres, the point of collision must lie on the line between the two centres. The relative velocity of the two molecules along this line of centres must be the sum of the individual components of velocity of the two molecules along this axis, in the sense that the component for A is measured left-to-right and that for B is measured right-to-left.

While the fraction outlined by the above premise is capable of being evaluated mathematically and this derivation may be consulted elsewhere*, the result so obtained may be reached by a route which involves less pages of mathematics. The thrust of the previous paragraph is that the desired fraction is that in which the sum of two specified components of velocity is at least a certain value. The equation is readily available for the distribution of velocities where the molecules may move only on a plane, that is in just two degrees of freedom. This equation, relating to one species moving in two dimensions, will ably deputise for the movement of two species each in one specified direction.

The equation for the fraction of all molecules, each of mass m, with a velocity in the permitted plane between r and $(r + dr)$ is:

* see, for example, Moore, J. W. and R. G. Pearson, 1981. *Kinetics and Mechanism*, 3rd edition. New York: Wiley, pp. 89–92.

$$\frac{\mathrm{d}n_r}{n} = \frac{m}{k_\mathrm{B}T} \exp\left(\frac{-mr^2}{2k_\mathrm{B}T}\right) r\,\mathrm{d}r \tag{4.7}$$

As it is more convenient to work in terms of kinetic energy, we equate ε to $\frac{1}{2}mr^2$ and $\mathrm{d}\varepsilon$ to $mr\mathrm{d}r$ and obtain, for the fraction of all molecules with a kinetic energy in these two degrees of freedom between ε and $(\varepsilon + \mathrm{d}\varepsilon)$:

$$\frac{\mathrm{d}n_\varepsilon}{n} = \frac{1}{k_\mathrm{B}T} \exp\left(\frac{-\varepsilon}{k_\mathrm{B}T}\right) \mathrm{d}\varepsilon \tag{4.8}$$

Thus the fraction, f, of all collisions which are sufficiently vigorous for reaction to be possible is given by

$$f = \int_{\varepsilon^*}^{\infty} \frac{\mathrm{d}n_\varepsilon}{n}$$

where ε^* represents the lower limit of feasibility. Thus we have:

$$\begin{aligned} f &= \int_{\varepsilon^*}^{\infty} \frac{1}{k_\mathrm{B}T} \exp\left(\frac{-\varepsilon}{k_\mathrm{B}T}\right) \mathrm{d}\varepsilon \\ &= \left[-\exp\left(\frac{-\varepsilon}{k_\mathrm{B}T}\right)\right]_{\varepsilon^*}^{\infty} = \exp\left(\frac{-\varepsilon^*}{k_\mathrm{B}T}\right) \end{aligned} \tag{4.9}$$

If numerator and denominator of the exponent are multiplied by Avogadro's constant, then f becomes $\exp(-E^*/RT)$.

There is, however, another consideration in regard to the efficiency of collisions between reactants. Whereas some reacting species are spherically symmetrical, others clearly are not and for some reactions it is obvious that many inter-reactant collisions, no matter how vigorous, must be unproductive simply because at the point of impact the reactants are oriented in an unfavourable way. An example is provided by a Diels–Alder reaction:

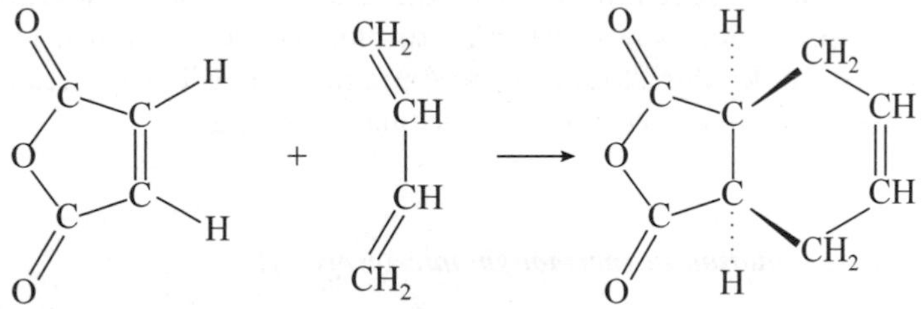

(maleic anhydride: O=C–O–C=O ring with HC=CH) + $CH_2{=}CH{-}CH{=}CH_2$ $\longrightarrow$ (cyclic adduct: anhydride ring fused to CH_2–CH=CH–CH_2 ring, with H at the ring-junction carbons) (4.10)

Unless these molecules collide in the appropriate orientation, reaction cannot be envisaged.

To meet this, it is normal to include in the expression a factor P, called the steric factor, which has a maximum value of one and in a case like this will be very much less than one. Thus the complete expression for the bimolecular rate constant provided by the Collision Theory is:

$$k = Pd^2\left[\frac{8\pi(M_A + M_B)RT}{M_A M_B}\right]^{1/2} \exp(-E^*/RT) \tag{4.11}$$

It has frequently been said that a great weakness of the Collision Theory is the lack of any means to evaluate P. This is true, though an order-of-magnitude assessment is feasible: for reaction (4.10), a value between 10^{-3} and 10^{-4} would seem to lie in the correct range.

A more serious weakness of the Collision Theory, if it is judged as a means of obtaining an *a priori* value for the rate constant of a reaction, is that there is no means by which E^* may be derived, estimated or guessed. Bearing in mind that if $E^* = 50\,\text{kJ mol}^{-1}$, the factor $\exp(-E^*/RT)$ is 1.72×10^{-9}, the seriousness of this deficiency is apparent. One approach is to borrow the answer provided by Transition State Theory for what, in the terms of that theory, would appear to be the corresponding parameter. Another is just to accept the experimental activation energy as being E^*.

4.2 Transition State Theory

The formulation of what is generally known as the Transition State Theory (TST) was achieved around 1935, with important contributions from a range of scientists including Evans, Eyring, M. Polanyi and Wigner. This has without doubt proved to be a much more useful theory of bimolecular reactions as is illustrated by these words of Lord Porter, spoken in 1962:

> Imperfect as it may be, [Transition State Theory] is undoubtedly the most useful theory that we possess. During the last 25 years, its greatest success has been, not in the accurate prediction of the rates of even the simplest reaction, but in providing a framework in terms of which even the most complex reaction can be better understood.

The usefulness of TST is not limited to any one phase, and it has proved extremely helpful for reactions both in the gas phase and in solution. In principle, its application is not restricted to bimolecular reactions. In practice, genuine termolecular reactions are extremely rare. Thus, virtually by default, it has become a theory of bimolecular reactions.

4.2.1 Simple formulation in thermodynamic terms

Transition state theory makes two essential postulates, which may be summarised as follows:

(i) In the course of the reaction of A with B, there will somewhere be a configuration of the nuclei of the constituent atoms such that the species so formed may *spontaneously* undergo reaction, *either* to the reaction products *or* back to the reactants, A and B.

(ii) This species, which is usually referred to as the *transition state* of the reaction, is assumed to be in effective equilibrium with the reactants.

These postulates may be summarised in the reaction scheme,

$$\mathrm{A + B} \quad \rightleftharpoons \quad \ddagger \rightarrow \text{ products} \tag{4.12}$$

where, following the established convention, the transition state is represented by the "double dagger" symbol, ‡. In the light of various schemes discussed in the previous chapter it is perhaps prudent to spell out just what this means.

Firstly, the transition state is envisaged as a totally unstable species, so that its concentration will always be negligible in relation to those of A and B. Thus the assumption that equilibrium is established does not require any allowance for the concomitant depletion of the concentrations of A and B. Secondly, the behaviour of the transition state differs from that of the intermediate complex in mechanism (3.15), in that the former remains in equilibrium with the reactant species, regardless of the rate at which reaction products are being formed.

For the first part of scheme (4.12) we can write the equilibrium constant, which we will call $K_{\ddagger}$.

$$K_{\ddagger} = \frac{[\ddagger]}{[\mathrm{A}][\mathrm{B}]} \tag{4.13}$$

The rate of the reaction, evaluated on the basis of the rate of the second step, whose first order rate constant we shall call $k_{\ddagger}$, is then given by:

$$\begin{aligned}\frac{\mathrm{d[Product]}}{\mathrm{d}t} &= k_{\ddagger}[\ddagger] \\ &= k_{\ddagger}K_{\ddagger}[\mathrm{A}][\mathrm{B}]\end{aligned} \tag{4.14}$$

The equilibrium constant for the first stage may of course be expressed in terms of the corresponding Gibbs energy change, $\Delta G^{\ominus}$.

$$\Delta G^{\ominus} = -RT\ln K_{\ddagger} \tag{4.15}$$

This parameter, known as the Gibbs energy of activation, is the difference between the standard Gibbs energy of the transition state and that of the reactants, A and B:

$$\Delta G^{\ominus}_{\ddagger} = G^{\ominus}_{\ddagger} - G^{\ominus}_{\mathrm{A+B}} \tag{4.16}$$

Thus the second order rate constant, which is the rate divided by the product of the concentrations of A and B, is given by:

$$k_2 = k_{\ddagger}\exp(-\Delta G^{\ominus}_{\ddagger}/RT) \tag{4.17}$$

Since $G = H - TS$, this may alternatively be expressed as:

$$k_2 = k_{\ddagger}\exp(\Delta S^{\ominus}_{\ddagger}/R)\exp(-\Delta H^{\ominus}_{\ddagger}/RT) \tag{4.17a}$$

4.2.2 Representation of the transition state using potential energy surfaces

If we consider a reaction in which an atom or a radical, X, is transferred from one group to another,

$$R{-}X + P^{\bullet} \rightarrow R^{\bullet} + X{-}P \tag{4.18}$$

then we may perceive that the potential energy of the system must vary with the inter-atomic distances and may be expressed in terms of two of them, $r_{R\text{–}X}$ and $r_{P\text{–}X}$, using contour lines to help in representing three parameters using only a two-dimensional sheet of paper.

A typical potential energy diagram for this system is sketched in Figure 4.2. When P is far removed from X (and from R), the variation of the potential energy with $r_{R\text{–}X}$ is similar to that of a diatomic molecule. Likewise, when $r_{R\text{–}X}$ is large, the potential energy shows a similar variation with $r_{P\text{–}X}$. When *P* approaches R–X (and here a linear configuration is assumed) the R–X bond is weakened and lengthens while incipient bonding occurs between P and X, leading to the configuration:

$$R\text{- - - - - -}X\text{- - - - - -}P \tag{4.19}$$

This means that there is a route by which reaction (4.18) may be achieved which does not involve making a total rupture of the R–X bond before

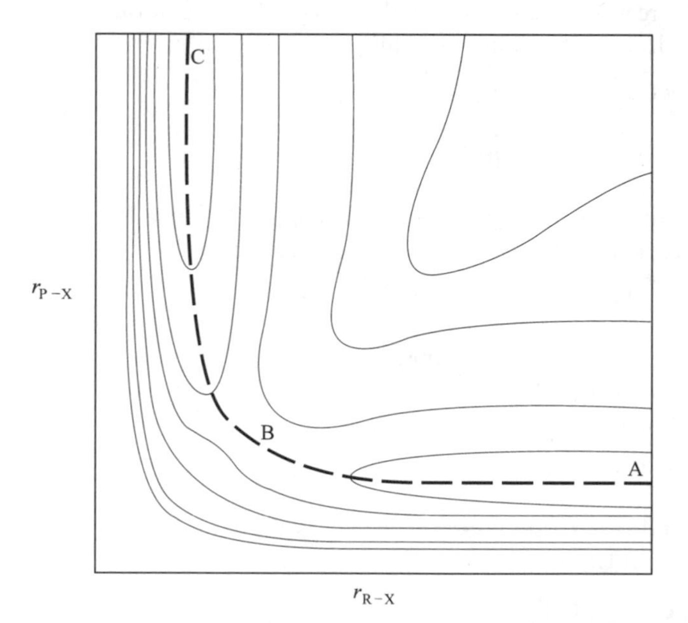

Fig. 4.2 Contours of potential energy relating to the distances, $r_{R\text{–}X}$ and $r_{P\text{–}X}$, for the interaction of the molecule RX with P to yield the reaction products R + PX. The dotted line (- - - - -) marks the easiest route by which this reaction might be achieved.

commencing to form the P–X bond. The pathway to achieving this is shown by the dashed line in Figure 4.2.

It may readily be seen that this dotted line starts, at A, in (to use geographical terms) a valley, and gains height as it rises up the valley until the point B is reached. Thereafter it loses height as it goes down another valley to C. In the terminology of a mountain track, B is thus a *col* or *saddle-point*: from B it is downwards towards A and towards C, but upwards in the other directions.

Sometimes, in referring to the change in potential energy along the reaction path, a graph is drawn which is essentially a plot of the potential energy along the dotted line AC, where the latter, which serves as the abscissa of the graph, is called the reaction co-ordinate. This is shown in Figure 4.3. The highest point on this curve represents the configuration of R- - -X- - -P such that either of the two processes is spontaneously possible. It thus represents the transition state of the reaction and is so labelled in Figure 4.3.

Considering R- - -X- - -P as a molecule, motion of X along the *R*- - - -P axis corresponds fairly closely to the asymmetric stretching vibration:

$$\overrightarrow{R}\text{- - - -}\overleftarrow{X}\text{- - - -}\overrightarrow{P}$$

However, the shape of Figure 4.3 is quite different from the potential energy curve for any molecular vibration, since it is concave downwards. Thus, firstly, it must be recognized that the transition state has to be treated as an anomalous molecule from which one vibrational mode is absent. Secondly, motion of X along the R- - -P axis is tantamount to movement along the reaction co-ordinate, and can thus shed light on the rate constant, $k_{\ddagger}$, for the conversion of the transition state to the reaction products.

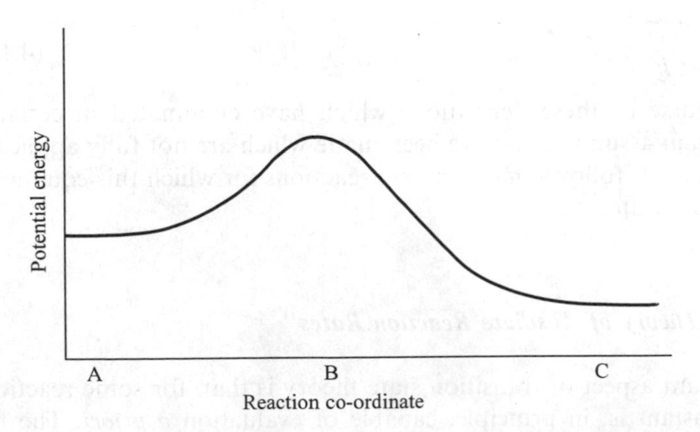

Fig. 4.3 Variation of potential energy with distance along the reaction co-ordinate, AC, using the data in Figure 4.2.

If motion along the reaction co-ordinate is treated as translational motion, by the Kinetic Theory its mean velocity will be given by

$$\bar{v} = \left(\frac{k_B T}{2\pi m^{\ddagger}}\right)^{1/2} \tag{4.20}$$

where $m^{\ddagger}$ represents the effective mass moving along the reaction co-ordinate. If the transition state, ‡, is regarded as extending a finite distance δ along the reaction co-ordinate, then the rate of passing through the transition state is given by $\bar{v}/\delta$.

The role of partition functions has not yet been introduced, but will be explained in the next section, where we will look at expressing the equilibrium constant $K_{\ddagger}$ in statistical thermodynamic terms using partition functions. Since motion along the reaction co-ordinate is not regarded as a mode belonging to the transition state, it is appropriate for the relevant partition function $q_t^{\ddagger}$, the translational partition function of the transition state, to be included here. Thus we have:

$$\begin{aligned} \text{Rate constant} &= \frac{\bar{v}}{\delta} \cdot q_t^{\ddagger} \\ &= \left(\frac{k_B T}{2\pi m^{\ddagger}}\right)^{1/2} \cdot \frac{1}{\delta} \cdot \frac{(2\pi m^{\ddagger} k_B T)^{1/2}\delta}{h} \\ &= \frac{k_B T}{h} \end{aligned} \tag{4.21}$$

On this basis the first order rate constant $k_{\ddagger}$ for the conversion of the transition state to the products has been identified as $k_B T/h$, where h is Planck's constant. However, it is usual to include also another factor, of value lying between the limits of 1 and 0, to take account of the non-formation of products due to some peculiarity of the potential energy surface. This is called κ, the transmission coefficient. Thus the bimolecular rate constant, in full, is:

$$k_2 = \kappa \frac{k_B T}{h} e^{-\Delta G_{\ddagger}^{\ominus}/RT} \tag{4.17b}$$

In the course of these derivations which have culminated in equation (4.17b), certain assumptions have been made which are not fully applicable to all reactions. It follows that there are reactions for which this equation is not totally apposite.

4.2.3 *The "Theory of Absolute Reaction Rates"*

One significant aspect of transition state theory is that, for some reactions, the rate constant is, in principle, capable of evaluation *a priori*. The first qualification is that we must restrict ourselves to reactions in the gas phase, the second that neither reactant is electrically charged and the third that, in the transition state, the plot of potential energy change along the reaction co-ordinate is concave downwards, as in Figure 4.3.

From equations (4.14) and (4.17b) we have, for the second order rate constant:

$$k_2 = \kappa \frac{k_B T}{h} K_\ddagger \tag{4.22}$$

In the gas phase it is easily possible to express an equilibrium constant in terms of the molecular partition functions and $\Delta\varepsilon^o$, representing the absolute difference between the lowest energy levels of the species on the two sides of the equation. Thus for reaction (4.18) we have,

$$K_\ddagger = \frac{q^o_\ddagger}{q^o_{RX} \cdot q^o_P} e^{-\Delta\varepsilon^o_\ddagger / k_B T} \tag{4.23}$$

where $K_\ddagger$ is the equilibrium constant in terms of concentrations and each q^o value is the molecular partition function per unit volume.

On the basis of the dictum that all of chemistry is implicit in the Schrödinger equation, $\Delta\epsilon_\ddagger$ is capable of evaluation. A description of the quantum mechanical techniques involved lies outside the scope of the present work. Basically, what is needed are calculations of the energy of the system at various values of $r_{R\text{–}X}$ and $r_{P\text{–}X}$, so as to map out the potential energy surface along the reaction co-ordinate. The difficulty in achieving high accuracy is in part due to the fact that what are being calculated are total interaction energies, using as reference state the separated nuclei and electrons. Thus $\Delta\varepsilon_\ddagger$ is evaluated as the small difference between substantial quantities, so that while each of these may be accurate to better than 1%, the error in $\Delta\varepsilon_\ddagger$ may be vastly greater than this. In this treatment, $\Delta\varepsilon_\ddagger$ will be considered as measured from the zero-point energies, i.e. $v = 0$, and not from the base of the potential energy curves.

However, experience with one of the simplest chemical reactions of all shows that accurate calculations of the potential energy surface are not easily achieved. For the reaction system, $H + H_2$, which involves only the minimum of three electrons, the early calculations (both semi-empirical and *ab initio*) yielded barrier heights somewhat at odds with experiment. Much better agreement has been achieved with this system since 1980, and the calculated values are now described as being as accurate as the experimental ones. However, for other reactions, such as $F + H_2$, the necessary calculations would be much more difficult in view of the greater number of electrons involved, so an accurate calculation for the barrier height is not anticipated.

The rate constant for reaction (4.16) will be at its simplest for the case where R, X and P are each atoms, when, additionally the calculation of $\Delta\varepsilon^o$ is least complex. Then the partition function for P will consist only of translational modes, the diatomic RX will have translational modes, one vibrational mode and two rotational modes, while the transition state,

R- - - -X- - - -P, being a linear triatomic, but special in the way that was discussed in Section 4.2.2, will have translational, three vibrational and two rotational modes.

In evaluating the partition functions for the rate constant of reaction (4.18), the only uncertainty arises from the unknown vibrational frequencies of the transition state. However, the form of the vibrational partition function, $q_{vib} = \{1 - \exp(-h\nu/k_B T)\}$, is such that, provided the vibrational frequency υ is reasonably large, q_{vib} is only slightly less than unity and so an intelligent guess for these vibrational frequencies (which are not accessible experimentally) can adequately deputise for the accurate value. So apart from that qualification to set beside the problem of accurately evaluating $\Delta\varepsilon_{\ddagger}$ (where an inaccuracy of 6 kJ mol^{-1} can cause an error of more than one order of magnitude), the *a priori* evaluation of the rate constant is feasible without the assistance of any kinetic experiments.

4.2.4 *A comparison of the Collision Theory and Transition State Theory*

To assist in a comparison of these two theories, it is instructive to expand equation (4.22) for the reaction of atoms A and B. Each reactant now has only translational energy modes while the transition state has also two rotational modes. This gives:

$$k_2 = \kappa\,\frac{k_B T}{h}\,\frac{[(2\pi(m_A + m_B)k_B T)]^{3/2}/h^3}{[(2\pi m_A k_B T)^{3/2}]/h^3}\,\frac{(8\pi^2 I k_B T)/h^2}{[(2\pi m_B k_B T)^{3/2}]/h^3}\,e^{-\Delta\varepsilon^0_{\ddagger}.k_B T} \tag{4.22a}$$

The moment of inertia I may be expressed in terms of the reduced mass and the separation l of the nuclei of A and B in the transition state:

$$I = \frac{m_A m_B}{m_A + m_B}\,l^2 \tag{4.24}$$

Equation (4.22a) may then be simplified:

$$\begin{aligned} k_2 &= \kappa\left\{\frac{8\pi k_B T(m_A + m_B)}{m_A m_B}\right\}^{1/2} l^2\, e^{-\Delta\varepsilon^0_{\ddagger}/k_B T} \\ &= \kappa\left\{\frac{8\pi RT(M_A + M_B)}{M_B M_B}\right\}^{1/2} l^2\, e^{-\Delta E^0_{\ddagger}/RT} \end{aligned} \tag{4.22b}$$

Comparison of the TST equation (4.22b) with the CT equation (4.11) reveals a remarkable similarity. The notable differences are: (i) TST has a factor κ, CT has a factor P; (ii) the TST equation includes the square of l, the inter-nuclear separation of A and B in the transition state where CT has the square of d, the collision diameter of A and B; and (iii) TST has $\Delta E_{\ddagger}$ where CT has E^*. In each instance, the two parameters are conceptually distinct.

The TST counterpart of the CT steric factor P is readily illustrated by considering what happens to equation (4.22) if the reactants A and B are assumed to be polyatomic. This requires the inclusion, in the equation, of the relevant partition functions for rotation and for vibration. Many vibrational frequencies of the reactant molecules are effectively unchanged in the transition state, so that these partition functions cancel out. Of the others, provided that the frequency is high, the vibrational partition function is approximately unity, so their inclusion will have little real effect.

Also, there will be additional factors for rotational modes, since each reactant and the transition state may have three rotational degrees of freedom. This entails the inclusion in equation (4.22) of the partition functions for one further rotational degree of freedom in the numerator and for six rotational degrees of freedom in the denominator. In consequence, the value for k_2 will be diminished by a factor of perhaps 10^4 or 10^5, dependent on the moments of inertia and the temperature. In this way TST provides a means, which is much less *ad hoc* than the P factor of CT, of estimating the factor by which geometrical considerations will attenuate the rate constant.

Another aspect of this same point is apparent if we consider a reversible bimolecular process, with rate constants k_f and k_r for the forward and reverse reactions:

$$\mathrm{A} + \mathrm{B} \rightleftharpoons \mathrm{C} + \mathrm{D} \tag{4.25}$$

The ratio of these rate constants is of course equal to the equilibrium constant, K:

$$K = \frac{[\mathrm{C}][\mathrm{D}]}{[\mathrm{A}][\mathrm{B}]} = \frac{k_f}{k_r} \tag{4.26}$$

If we take for these rate constants the respective Collision Theory (CT) expressions, we have:

$$\begin{aligned} K &= \frac{P_{AB} Z_{AB}\, e^{-E^*_{AB}/RT}}{P_{CD} Z_{CD}\, e^{-E^*_{CD}/RT}} \\ &= \frac{P_{AB}}{P_{CD}} \cdot \frac{Z_{AB}}{Z_{CD}} \cdot e^{-(E^*_{AB} - E^*_{CD})/RT} \end{aligned} \tag{4.26a}$$

where P_{AB} and P_{CD} are the steric factors of the forward and reverse reactions, whose collision numbers are Z_{AB} and Z_{CD} and whose minimum kinetic energy requirements along the line of centres are E^*_{AB} *and* E^*_{CD}.

This equation is in contrast to that derived from thermodynamics,

$$K = e^{-\Delta G^\ominus/RT} = e^{\Delta S^\ominus/R} \cdot e^{-\Delta H^\ominus/RT} \tag{4.27}$$

in that equation (4.26a) seems to contain no entropy term. The corresponding quotient using the ratio of the TST rate equations contains the requisite entropy term and in this respect the TST approach is more satisfactory than that of CT.

It is also appropriate to look at the relationship between the CT and the TST equations for a second order rate constant and the empirical Arrhenius equation. To recap, these are as follows:

$$\text{CT} \qquad k_2 = Pd^2\left[\frac{8\pi(M_A + M_B)RT}{M_A M_B}\right]^{1/2} e^{-E^*/RT} \tag{4.11}$$

$$\text{TST} \qquad k_2 = \kappa\,\frac{k_B T}{h}\, e^{\Delta S^{\ominus}_{\ddagger}/R}\, e^{-\Delta H^{\ominus}_{\ddagger}/RT} \tag{4.17b}$$

$$\text{Arrhenius} \quad k_2 = A\, e^{-E_a/RT} \tag{1.44}$$

All three equations have an exponential factor. In CT, the pre-exponential factor is proportional to $T^{1/2}$, in TST, it is proportional to T and in the Arrhenius equation it is independent of T. In consequence, the Arrhenius activation energy E_a — which is equal to $-R\,\mathrm{d}\ln k/\mathrm{d}(1/T)$ — corresponds to $(E^* + \frac{1}{2}RT)$ or to $(\Delta H^{\ominus}_{\ddagger} + RT)$. As a result, the Arrhenius A factor corresponds as follows:

$$\text{CT} \quad A \equiv Pd^2\left[\frac{8\pi(M_A + M_B)RT}{M_A M_B}\right]^{1/2} e^{1/2} \tag{4.28}$$

$$\text{TST} \quad A \equiv \kappa\,\frac{k_B T}{h}\, e^{\Delta S^{\ominus}_{\ddagger}/RT} \cdot e \tag{4.29}$$

4.3 The theory of diffusion-controlled reactions in solution

One group of bimolecular reactions whose study is not materially assisted by the Transition State Theory is that of fast reactions in solution. The fundamental reason is that if reaction should take place on virtually every collision between the two reactant species, the real difficulty to be surmounted to achieve reaction is not the formation of the transition state but rather the mutual approach of the reactants through the solvent. Thus the reactants and the transition state are effectively not in equilibrium and the concentration of the transition state is, because of transport constraints, substantially less than $K_{\ddagger}$[A][B].

Regarding the possible relevance of Collision Theory to such reactions, as was mentioned earlier, the rate of A–B collisions in solution is very much the same as it is in the gas phase. However, the pattern of these collisions is quite different. In the gas phase, the collisions experienced by an A molecule with a B molecule are randomly spaced and, almost without exception, each successive collision is with a different B molecule. This follows naturally because, even at moderate pressures, the mean free path of a molecule is hundreds or thousands of times greater than the molecular diameter.

On the other hand, when an A–B collision occurs in solution, the same A and B molecules, held together by a cage of solvent molecules, undergo many collisions in close succession before they can escape from each other. This occurrence is sometimes termed an encounter, and the number of A–B collisions within it, while a function of such variables as the solvent and the temperature, may lie in the range 5 to 25. If the reaction between A and B is slow, so that only one collision in 10^4 or in 10^6 leads to reaction, the nature of this pattern is of no significance. But if A and B react so readily that reaction occurs on the first collision, then it is not the (hypothetical) total number of collisions that matters, but only the first collision of each encounter.

The best theoretical approach to reactions of this kind is one which addresses the rate-limiting process, the diffusion of the reactants through the solvent towards each other.

4.3.1 The diffusion model of Smoluchowski

A theory of reactions in solution, occurring in the first collision of the reacting species, was proposed by Smoluchowski in 1917. The actual process he was concerned with was the coagulation of colloidal particles, which is also controlled by diffusion, but the model is applicable also to purely chemical processes.

If molecules A and B react on their first collision it is useful to represent a molecule of A by a spherical sink and assume that once the centre of a molecule of B enters this sphere, it disappears. The radius of the sink should thus be the sum of the radii of the two species, i.e. $(r_A + r_B) = \sigma$. Since the sink will be used as our point of reference, there must be attributed to each molecule of B the correct diffusion coefficient, namely, the diffusion coefficient of B with respect to A, D_{AB}, which is known to be simply the sum of the individual diffusion coefficients, i.e. $D_{AB} = (D_A + D_B)$.

The steady state situation will involve the diffusion of molecules of B towards the sink, in such a way that at any point the concentration c of these molecules stays constant with time. To this model we may apply Fick's First Law of Diffusion, which states that in the steady state the net movement of diffusing molecules per unit area is proportional to the concentration gradient and to the diffusion coefficient. Thus at a distance r (where r is greater than σ) from the centre of the sink, the flux J of molecules of B crossing the surface of this sphere towards the sink will be given by:

$$\left(\frac{\mathrm{d}c}{\mathrm{d}r}\right)_r = \frac{J}{4\pi r^2 D_{AB}} \tag{4.30}$$

If a steady state situation prevails, then the value of J does not depend on r, so in equation (4.30) there are only two variables, c and r. Separating these and integrating we obtain:

$$c_r = -\frac{J}{4\pi r D_{AB}} + \text{const} \tag{4.31}$$

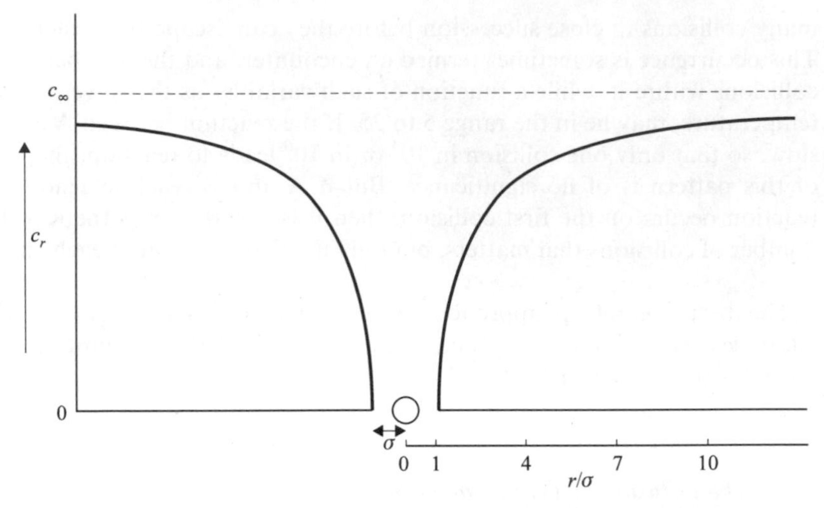

Fig. 4.4 The variation of the concentration, c_r, of the diffusing species around the stationary sink of the Smoluchowski diffusion model. The dashed line designates c_∞.

The constant of integration may be evaluated if we assume that at very large distances from the sink, the concentration of B molecules is simply c_∞, the overall mean concentration of B in the solution. This leads to:

$$c_r = c_\infty - \frac{J}{4\pi r D_{AB}} \tag{4.32}$$

For the lower boundary condition, Smoluchowski assumed that at $r = \sigma$, c became zero. Thus, around a sink, the time-average concentration, c_r, of species B was envisaged to vary with r as shown in Figure 4.4. Also, by putting $r = \sigma$ we have:

$$c_\infty = \frac{J}{4\pi\sigma D_{AB}}$$

or

$$J = 4\pi\sigma D_{AB} c_\infty \tag{4.33}$$

Having thus, using the laws of diffusion, evaluated the flux of molecules B towards the sink which represents a reactive species A, we may obtain an expression for the second order rate constant k_2 which should characterise the rate of reaction of A with B. To obtain the total rate of reaction, we take the product of J (the flux towards one sink) and c_A, the concentration of A in the solution, which represents the concentration of such sinks. Dividing this by the product of the concentrations of A and of B,

$$k_2 = \frac{Jc_A}{c_\infty c_A} = 4\pi\sigma D_{AB} \tag{4.34}$$

we have Smoluchowski's expression for the second order rate constant of this diffusion-controlled reaction, occurring on the first collision of the species A and B.

In this derivation, it has been convenient to express concentrations as molecules per unit volume. Multiplying by Avogadro's constant, we have,

$$k_2/\mathrm{m^3\,mol^{-1}\,s^{-1}} = 4\pi\sigma D_{AB} N_A \tag{4.34a}$$

where σ and D_{AB} are both in SI units.

A typical value is obtained for the rate constant of a diffusion-controlled reaction by taking $\sigma = 0.3\,\mathrm{nm} = 3 \times 10^{-10}\,\mathrm{m}$ and $D_{AB} = 4 \times 10^{-9}\,\mathrm{m^2\,s^{-1}}$. This yields $k_2 = 9.1 \times 10^9\,\mathrm{dm^3\,mol^{-1}\,s^{-1}}$, which is less by just over an order of magnitude than the bimolecular collision number evaluated above from equation (4.6b), which gave $1.14 \times 10^{11}\,\mathrm{dm^3\,mol^{-1}\,s^{-1}}$. This comparison would support the contention that in solution the average number of collisions per encounter may be in the range 10 to 15.

Alternatively, this equation for the rate constant of a diffusion-controlled reaction may be simplified by using the Stokes–Einstein equation to express a diffusion coefficient in terms of the radius of the species concerned and the viscosity coefficient η of the solvent:

$$D_A = \frac{k_B T}{6\pi r_A \eta}$$

Substituting this into equation (4.34a), we obtain,

$$\begin{aligned} k_2/\mathrm{m^3\,mol^{-1}\,s^{-1}} &= 4\pi(r_A + r_B)\frac{k_B T}{6\pi\eta}\left(\frac{1}{r_A} + \frac{1}{r_B}\right)N_A \\ &\approx \frac{8RT}{3\eta} \end{aligned} \tag{4.34b}$$

where the relation is exact where $r_A = r_B$. Using the viscosity coefficient for water at 298 K of $0.890 \times 10^{-3}\,\mathrm{kg\,m^{-1}\,s^{-1}}$, the diffusion controlled rate constant for a reaction in aqueous solution is estimated as $7.4 \times 10^9\,\mathrm{dm^3\,mol^{-1}\,s^{-1}}$.

4.3.2 Reactions that are nearly diffusion controlled

Whereas colloidal particles might be expected always to coalesce on making contact, our reactants A and B could conceivably collide and not react. To cope with such a regime, a different lower boundary condition has been proposed as an alternative to that used by Smoluchowski.

By the radiation boundary condition, as this is called, the concentration of the diffusing species at $r = \sigma$ is taken as finite, and the steady state flux J is equated to kc_σ, where k denotes the (hypothetical) rate constant that would be applicable if the concentration of B were maintained uniform. Thus from equation (4.32) we obtain,

$$c_\sigma = c_\infty - \frac{kc_\sigma}{4\pi\sigma D_{AB}} \tag{4.35}$$

which leads to:

$$c_\sigma = \frac{c_\infty}{1 + (k/4\pi\sigma D_{AB})} \tag{4.36}$$

However, it is the operative second order rate constant which is experimentally accessible. As before, this is equal to J/c_∞, so we now have,

$$k_2 = \frac{kc_\sigma}{c_\infty} = \frac{4\pi\sigma D_{AB}}{1 + (4\pi\sigma D_{AB}/k)} \tag{4.37}$$

If the reaction is diffusion controlled, $k \gg 4\pi\sigma D_{AB}$, so that equation (4.37) scarcely differs from equation (4.34). If reaction occurs on only one collision in 10^4, or one encounter in *ca* 10^3, then $k \ll 4\pi\sigma D_{AB}$ so that $k_2 \approx k$. The merit of equation (4.37) is that it is applicable to the region in between these extreme situations.

Where reaction occurs on the first collision between the reactants, $k_2 = 4\pi\sigma D_{AB}$, and the only factor here that is temperature-dependent is the diffusion coefficient. A diffusion coefficient varies with temperature as $D = D_o \exp(-E_D/RT)$, so the Arrhenius activation energy of a diffusion-controlled reaction will be close to E_D.

Rearranging equation (4.37) we have,

$$\frac{1}{k_2} = \frac{1}{4\pi\sigma D_{AB}} + \frac{1}{k} \tag{4.38}$$

which may be used to estimate values of k_2 and thus to derive values for its Arrhenius activation energy.

There tends to be more data on the temperature dependence of the coefficient of viscosity, η, than on that of diffusion coefficients. For a "Stokes" fluid, the two parameters are related in that $D\eta/T$ is a constant. Moreover, on the basis of its measured coefficient of self-diffusion, it seems that this relation is applicable to water. Equating the value of $4\pi\sigma D_{AB}$ at 298 K to 7.5×10^9 dm^3 mol^{-1} s^{-1}, those at 278 K and at 323 K were obtained using the above relationship and the viscosities of water at these temperatures.

The hypothetical rate constant k was taken as $Z\exp(-E^*/RT)$, where Z was assumed to be 1.15×10^{11} dm^3 mol^{-1} s^{-1} at 298 K and to vary at $T^{1/2}$. On that basis, for various assumed values of E^*, starting at zero, the Arrhenius activation energy of k_2 was calculated, as is shown in Figure 4.5. This shows that the lowest Arrhenius activation energy occurs, not when E^* is zero (that is, for a reaction occurring on every collision), but rather when E^* is finite but quite small. The value of E^* at which the minimum occurs depends on the particular assumptions made regarding the relative sizes of $4\pi\sigma D_{AB}$ and Z.

The Arrhenius activation energy at $E^* = 0$ is 18.3 kJ mol^{-1}, which is just slightly less than the value of $(B + RT)$, where B is the parameter in the empirical equation, $\eta = A\exp(B/RT)$, for the temperature dependence of

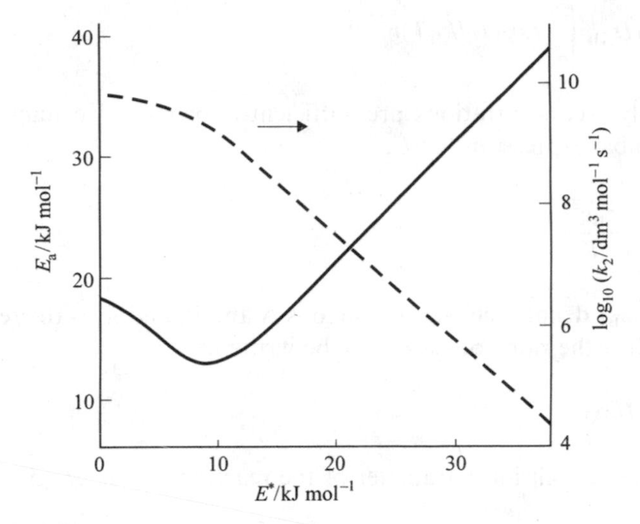

Fig. 4.5 Illustration of the significance of equation (4.32) for a reaction which occurs on not quite the first collision, Plot (—) and l.h. scale, of E_a against E^*, the parameter which determines the value of k, and, r.h. scale, of $\log_{10} k_2$ (- - -). The data refer to a reaction in aqueous solution, and the values of the various parameters are detailed in the text

the coefficient of viscosity. After decreasing to just under $13\,\mathrm{kJ\,mol^{-1}}$, E_a then starts to rise and, above $E^* = 16\,\mathrm{kJ\,mol^{-1}}$, it remains equal to $(E^* + \frac{1}{2}RT)$.

In using equation (4.37), the value of k, the rate constant that would be applicable if concentrations were maintained uniform, might be assessed either on the basis of the Collision Theory as has been done above, or on that of Transition State Theory. The nature of the diffusion model employed to treat fast reactions in solution in no way imposes any constraints on how the value of k should be estimated.

4.3.3 Debye's extension to reactions involving ions

An extension to the Smoluchowski theory was developed by Debye to include the effects of long range Coulombic forces between the reactants when both are ionic. To take account of this perturbation to the diffusive motion he added another term to equation (4.30), giving,

$$\frac{\mathrm{d}c}{\mathrm{d}r} + \frac{c}{k_\mathrm{B}T} \cdot \frac{\mathrm{d}U}{\mathrm{d}r} = \frac{J}{4\pi r^2 D_\mathrm{AB}} \qquad (4.39)$$

where $U(r)$ denotes the potential energy when A and B are at a separation r. This leads, using Smoluchowski's boundary conditions, to the equation:

$$k_2 = 4\pi\sigma D_{AB}\int_{\sigma}^{\infty}\exp(U/k_BT)r^{-2}\,dr \tag{4.40}$$

Where electrolyte concentrations are sufficiently low that we may use a simple Coulombic expression for U,

$$U(r) = \frac{z_A z_B e^2}{4\pi\varepsilon_o K_r r} \tag{4.41}$$

where $z_A e$ and $z_B e$ denote the charges on ions A and B, and K_r is the relative permittivity, then the rate constant may be written,

$$k_2 = 4\pi\sigma f D_{AB} \tag{4.42}$$

where the effective collision parameter of the reacting ions is σf, given by:

$$\sigma f = \frac{z_A z_B e^2}{4\pi\varepsilon_o K_r k_B T}\Bigg/\left\{\exp\left(\frac{z_A z_B e^2}{4\pi\varepsilon_o K_r \sigma k_B T}\right) - 1\right\} \tag{4.43}$$

This gives $f > 1$ when $z_A z_B$ is negative and $f < 1$ when $z_A z_B$ is positive, as would be expected since ions of unlike sign mutually attract and those of like sign mutually repel.

Suggested Reading

Greene, E. F. and A. Kuppermann, 1968. Chemical reaction cross-sections and rate constants. *Journal of Chemical Education,* **45**, 361–9.

Moore, J. W. and R. G. Pearson, 1981. *Kinetics and Mechanism*, 3rd edition. New York: Wiley.

McAlduff, E. J., 1980. An introduction to collision theory rate constants via distribution functions. *Journal of Chemical Education*, **57**, 627–8.

Nordman, C. E. and S. M. Blinder, 1975. Collision Theory of chemical reactions. *Journal of Chemical Education,* **51**, 790–1.

Laidler, K. J. and M. C. King, 1983. The development of transition-state theory. *Journal of Physical Chemistry,* **87**, 2657–64.

Mahan, B. H., 1974. Activated complex theory of bimolecular reactions, *Journal of Chemical Education,* **51**, 709–11.

Moss, S. J. and C. J. Coady, 1983. Potential energy surfaces and transition state theory. *Journal of Chemical Education,* **60,** 455–61.

Noyes, R. M., 1961. Effects of diffusion rates on chemical kinetics, *Progress in Reaction Kinetics,* **1**, 129–60.

Logan, S. R., 1967. Effects of temperature on the rates of diffusion-controlled reactions. *Transactions of the Faraday Society*, **63**, 1712–19.

Albery, W. J., 1993. Transition state theory revisited. *Advances in Physical Organic Chemistry,* **28**, 139–70.

Problems

4.1 A vessel contains O_2 at 10 torr pressure and a temperature of 298 K. Assuming a molecular diameter of 0.295 nm, how many O_2–O_2 collisions will occur per cm^3 per second?

4.2 Species A and B react together in the gas phase. The relative molar masses are $M_A = 0.098$ and $M_B = 0.054\,kg\,mol^{-1}$. If the collision diameter $\sigma = 0.4\,nm$, the steric factor $P = 0.0005$ and $E^* = 31.5\,kJ\,mol^{-1}$, evaluate the anticipated second order rate constant at 353 K.

4.3 What value of the entropy of activation of a bimolecular reaction has the same effect on the pre-exponential factor of the second order rate constant as a steric factor of 10^{-5} ?

4.4 The rate constant for the recombination of iodine atoms in solution, $I + I \rightarrow I_2$, is given by the Smoluchowski equation, $2k_2 = 4\pi\sigma D_{AA} N_A$. Using the Stokes–Einstein relation, $D_A = k_B T / 6\pi\eta r_A$, to relate the diffusion coefficient D_A to the radius r_A of the diffusing species and the viscosity η of the solvent, estimate the rate constant of this reaction at 298 K in *n*-hexane, for which $\eta = 2.94 \times 10^{-4}\,kg\,m^{-1}\,s^{-1}$ at this temperature.

4.5 For a reaction between ions of opposite charge, the Smoluchowski–Debye theory yields a minimum value of the bimolecular rate constant, regardless of the collision diameter. What is the minimum value of the second order rate constant for a diffusion-controlled reaction in aqueous solution at 298 K between a uni-negative ion and a divalent cation whose diffusion coefficients add up to $6 \times 10^{-9}\,m^2\,s^{-1}$?

4.6 At 298 K, the second order rate equation for the reaction, $H^+ + OH^-$, is $1.35 \times 10^{11}\,dm^3\,mol^{-1}\,s^{-1}$. Given that the sum of the diffusion coefficients of these species is $14.6 \times 10^{-9}\,m^2\,s^{-1}$, what value of the collision diameter σ is required in the Smoluchowski–Debye equation in order to accommodate such a high rate constant?

5 The interpretation of bimolecular reactions in solution

The chief role of the Transition State Theory lies in providing a logical framework on the basis of which it is possible to interpret kinetic measurements on bimolecular reactions. This is clearly illustrated by reactions in solution and in this chapter we consider kinetic effects due to the solvent, to applied hydrostatic pressure and to non-participating added salts. There are also introductions to linear free energy relationships and the Hammett equation, to kinetic isotope effects and to the Marcus theory of electron transfer in solution.

5.1 Solvent effects on reaction rates

From Chapter 4 we have the basic TST relation, known as the Eyring equation, where the rate constant k_2 is expressed in terms of thermodynamic parameters relating to the transition state:

$$\begin{aligned} k_2 &= \frac{k_B T}{h} \exp\left(\frac{-\Delta G_\ddagger^\ominus}{RT}\right) \\ &= \frac{k_B T}{h} \exp\left(\frac{\Delta S_\ddagger^\ominus}{R}\right) \exp\left(\frac{-\Delta H_\ddagger^\ominus}{RT}\right) \end{aligned} \tag{5.1}$$

If the rate constant of a reaction has been determined as a function of temperature then, assuming that the standard entropy of activation $\Delta S_\ddagger^\ominus$ and the standard enthalpy of activation $\Delta H_\ddagger^\ominus$ are both independent of temperature, these two quantities may readily be determined from the data. Rearranging equation (5.1) and taking logarithms of both sides of the equation we have:

$$\ln\left(\frac{k_2}{T}\right) = \ln\left(\frac{k_B}{h}\right) + \frac{\Delta S_\ddagger^\ominus}{R} - \frac{\Delta H_\ddagger^\ominus}{RT} \tag{5.2}$$

Thus ln (k_2/T) is expected to be a linear function of T^{-1}, and this graph is known as an Eyring plot. The slope is equal to $-\Delta H_\ddagger^\ominus/R$ and once the standard enthalpy of activation has been determined, the standard entropy of activation $\Delta S_\ddagger^\ominus$ is readily found by substitution in equation (5.2).

As explained in Chapter 4, an activation parameter is the difference between the value of that property for the transition state and the value for the reactants, A and B. For a reaction in solution, the solvent is involved to an appreciable extent with both these entities, so that the activation parameter,

$$\Delta X_{\ddagger} = X_{\ddagger(\text{solvated})} - X_{\text{A+B(solvated)}} \qquad (5.3)$$

reflects the change in the behaviour of the solvent in the course of forming the transition state, ‡, from the reactants.

Where the transition state is more highly polarised than the reacting species, as is the case in the reaction,

$$R_3N + QI \rightleftharpoons R_3\overset{\delta+}{N} \ldots Q \ldots \overset{\delta-}{I} \longrightarrow R_3Q\overset{+}{N} + I^- \qquad (5.4)$$

then one would expect that, in a polar medium, the transition state should be more strongly solvated than the reactants. This means that the solvent molecules will be more highly oriented around the transition state than around the reactants. Consequently, the formation of the transition state will be accompanied by an appreciable loss of entropy, so that $\Delta S^{\ominus}_{\ddagger}$ will be highly negative.

Another consequence of the strong solvation by a polar solvent is that, for reaction (5.4), the enthalpy of the strongly polar transition state is lowered to a greater extent than is that of the reactant species. Thus the enthalpy of activation is lowered relative to the value in a non-polar solvent. For example, for the reaction of EtI with Et_3N, it is reported that $\Delta H^{\ominus}_{\ddagger}$ falls from 64 kJ mol^{-1} in *n*-hexane to 49.5 kJ mol^{-1} in bromobenzene and 45.5 kJ mol^{-1} in nitrobenzene.

If, on the other hand, the transition state is less strongly solvated than are the reactants, then the opposite effect is to be expected. For example, for the reaction, in aqueous solution,

$$[Cr(H_2O)_6]^{3+} + SCN^- \rightarrow [Cr(H_2O)_5SCN]^{2+} + H_2O \qquad (5.5)$$

$\Delta S^{\ominus}_{\ddagger}$ has the value +110 kJ mol^{-1}, which is a reflection of the extent to which solvent molecules are less tightly bound around the transition state than they are around the reactant ions.

5.2 Applied hydrostatic pressure

The application of hydrostatic pressure to a solution in which a chemical reaction is taking place normally alters the reaction rate. This change, which is sometimes an acceleration and sometimes a retardation, may be orders of magnitude greater than the shrinkage in volume which accompanies the application of pressure to the solution. In fact, this slight increase in reactant concentrations is totally incidental to the effect.

For an understanding of this phenomenon, let us start from equation (5.1) and take logarithms of both sides:

$$\ln k_2 = \ln \frac{k_B T}{h} - \frac{\Delta G^{\ominus}_{\ddagger}}{RT} \tag{5.6}$$

On the right hand side of this equation, the value of the first term is invariant at constant temperature since the factors other than T are fundamental constants. To deduce the pressure dependence of the second term, we may start from the basic expression for dG for a closed system:

$$\mathrm{d}G = V.\mathrm{d}P - S.\mathrm{d}T \tag{5.7}$$

From this it follows that:

$$\left(\frac{\partial G}{\partial P}\right)_T = V \tag{5.8}$$

Thus when equation (5.6) is differentiated with respect to pressure at constant temperature we have:

$$\left(\frac{\partial \ln k_2}{\partial P}\right)_T = \frac{-\Delta V_{\ddagger}}{RT} \tag{5.9}$$

where $\Delta V_{\ddagger}$ is the activation volume of the reaction.

Equation (5.9) implies that whether the gradient of ln k_2 against P is positive or negative depends on whether $\Delta V_{\ddagger}$ is negative or positive. Also, provided the activation volume does not vary with applied pressure, then such a plot is predicted to be a straight line. In general, these plots show slight curvature with the gradient becoming arithmetically less at increased pressures. This is consistent with the difference between $V_{\ddagger(\mathrm{solvated})}$ and $V_{\mathrm{A+B(solvated)}}$ becoming progressively smaller as the applied pressure is increased.

Some examples of the effect of hydrostatic pressure on the rate constant of bimolecular reactions in solution are shown in Figure 5.1. In this selection, applied pressure in each case causes an acceleration; however, there are many instances of the opposite effect, as is illustrated below.

The limiting value of the activation volume is readily derived from the initial gradient. In this evaluation it is convenient to express all parameters in SI units. For reaction (c), namely,

$$CH_3CO_2CH_3 + OH^- \rightarrow CH_3CO_2^- + CH_3OH \tag{5.10}$$

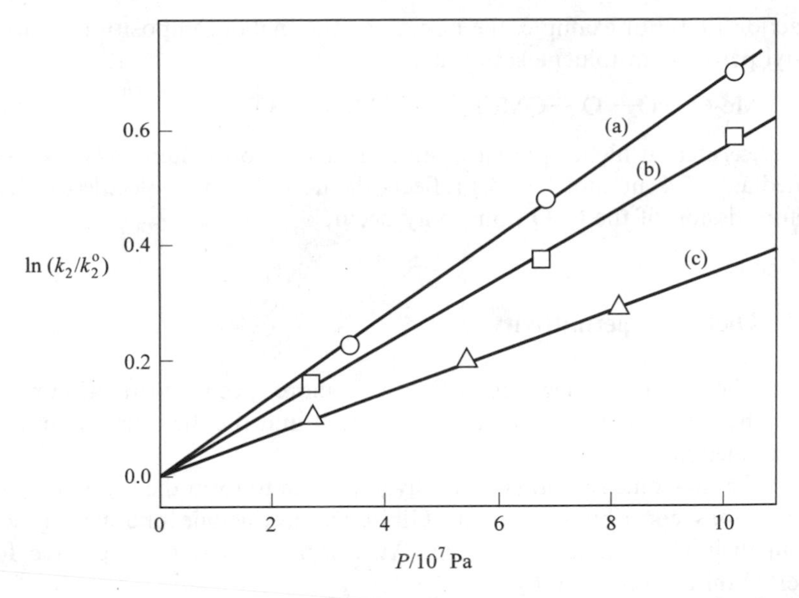

Fig. 5.1 Plot of $\ln(k_2/k_2^o)$ against applied hydrostatic pressure P for the hydrolysis reactions: (a) $C_2H_5CONH_2$ + OH^-; (b) CH_3CONH_2 + OH^-; and (c) $CH_3CO_2C_2H_5$ + OH^-. (Data from Laidler, K. J. and D. Chen, *Trans Faraday Soc.*, 1958, **54**, 1026–33).

we may evaluate $\Delta V_{\ddagger}$ as follows:

$$\begin{aligned}\Delta V_{\ddagger} &= -RT\left(\frac{\partial \ln k_2}{\partial P}\right)_T \\ &= -8.314 \times 298.1 \times \frac{0.24}{6.00 \times 10^7} \\ &= -9.9 \times 10^{-6}\,\text{m}^3\,\text{mol}^{-1} \\ &= -9.9\,\text{cm}^3\,\text{mol}^{-1}\end{aligned} \tag{5.11}$$

For a reaction such as (5.10), involving ions in a polar solvent, there is a general correlation between the sign and magnitude of $\Delta V_{\ddagger}$ and $\Delta S_{\ddagger}^{\ominus}$, arising from the involvement of the solvent. Where the transition state is more polar than the reactants, and thus is more highly solvated, the solvent molecules, tightly held around the transition state, will have a lower entropy and will occupy a lesser volume than the same number of solvent molecules in the neighbourhood of the reactants. Where the *TS* is less polar than the reactants, then both the entropy and the volume of the solvent will increase around the transition state, so that both $\Delta S_{\ddagger}^{\ominus}$ and $\Delta V_{\ddagger}$ are positive quantities.

However, polarity effects are not always dominant in determining the activation volume. For reactions in non-polar solvents not involving ions, solvent effects are minimal, but applied pressure may appreciably affect the reaction rate. For example, the homolytic thermal decomposition of di-*tert*-butyl peroxide in toluene solution,

$$Me_3C—O—O—CMe_3 \rightarrow 2\,Me_3C—O^{\cdot} \tag{5.12}$$

is decelerated at higher pressures and the activation volume has been evaluated as $+5.4\,cm^3\,mol^{-1}$. This reflects the need for the molecule to expand before fission of the O–O bond may occur.

5.3 Dielectric permittivity

When the reacting species are both ions, a more specific correlation may be made between the bimolecular rate constant and the dielectric permittivity of the medium.

As the two ionic reactants mutually approach to form the transition state, the various contributions to the Gibbs energy include one arising solely from their Coulombic interaction. At a separation r, the repulsive force exerted on each is given by:

$$F = \frac{z_A z_B e^2}{4\pi\varepsilon_o K_r r^2} \tag{5.13}$$

where K_r denotes the relative permittivity and $z_A e$ and $z_B e$ the charges on ions A and B. Thus the Coulombic potential energy of ions A and B at separation r is given by:

$$E = \frac{z_A z_B e^2}{4\pi\varepsilon_o K_r r} \tag{5.14}$$

The Gibbs energy of activation may be written as the sum of two terms:

$$\Delta G^{\ominus}_{\ddagger} = \Delta G^{\ominus}_{\ddagger}\,(\text{non-Coul}) + \Delta G^{\ominus}_{\ddagger}\,(\text{Coul}) \tag{5.15}$$

The Coulombic term in equation (5.15) will be well approximated by Avogadro's constant times the value of E when $r = d$, the separation of the charges of A and B in the transition state. On substituting equation (5.15) into equation (5.1) and taking logarithms we obtain as follows:

$$\ln k_2 = \ln\left(\frac{k_B T}{h}\right) - \frac{\Delta G^{\ominus}_{\ddagger}\,(\text{non-Coul})}{RT} - \frac{z_A z_B e^2 N_A}{4\pi\varepsilon_o K_r d RT} \tag{5.16}$$

This shows that, provided the variation in K_r is achieved without substantially altering any other parameter, then at constant temperature the logarithm of the rate constant should be a linear function of K_r^{-1}, with a gradient proportional to $-z_A z_B/d$.

Variations in the relative permittivity may of course bring about changes in the Gibbs energy of activation, additional to those detailed by the Coulombic term in equation (5.16). It is desirable that the physical nature of the solvent is not varied: a typical instance where equation (5.16) might be

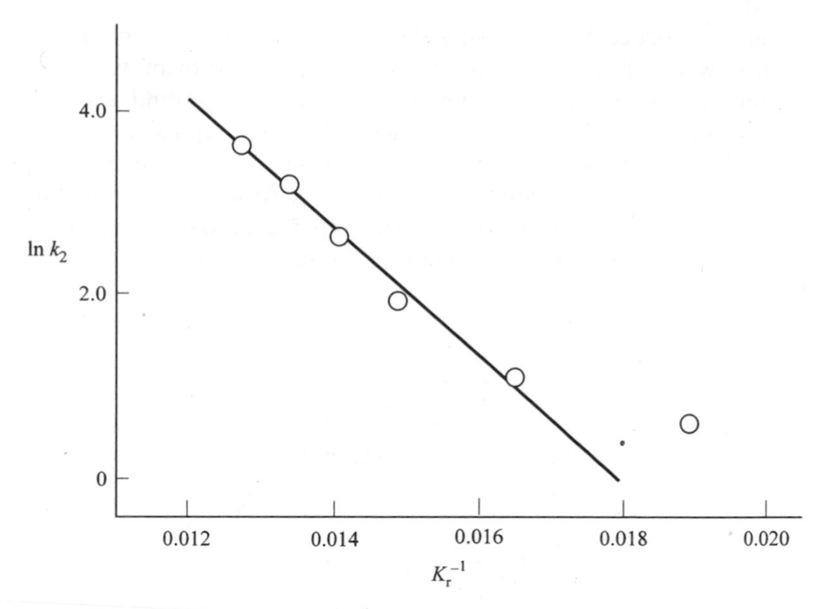

Fig. 5.2 Plot of ln k_2 against K_r^{-1} for the reaction of bromophenol blue with OH^- in water/methanol mixtures. For this reaction, the product of the charges on the participating ions is +2, and from the slope, d is evaluated as 0.16 nm. (Data from Amis, E. S. and V. K. LaMer, *J. Am. Chem. Soc.*, 1939, **61**, 905–13.)

deemed applicable would involve the use of a series of binary mixtures of the same two solvents, where K_r varies with composition. An example of such an application is shown in Figure 5.2.

5.4 Ionic strength

When a bimolecular reaction between ionic species takes place in a medium of high dielectric permittivity, such as water, then the value of the rate constant is found to depend on the ionic strength of the solution. This is sometimes called the primary kinetic salt effect.

The direction of this effect may readily be deduced from a qualitative consideration of the Debye–Hückel theory of electrolytical solutions. This shows that around any ion in solution there will be a preponderance of ions of opposite sign, known as the ionic atmosphere, and also that the characteristic radius of the ionic atmosphere, κ, decreases with increase in the ionic strength of the solution. When a pair of reactant ions, A and B, are at a separation significantly greater than κ, then the charge of each is entirely screened from the other by the ionic atmospheres. If, on the other hand, the separation is very much less than κ, then the interaction will be essentially Coulombic. It follows that, for a pair of ions of unlike sign, their reaction should be retarded by increasing the ionic strength, since this reduces the extent to which the approach of the ions is accelerated by mutual attraction. For ions of like sign, increasing the ionic strength should accelerate the reaction because it diminishes the extent of mutual repulsion.

The theoretical treatment of the primary salt effect stems from the fact that whereas $K_{\ddagger}$, the equilibrium constant for forming the transition state from the reactants, is a function only of temperature, the activity coefficients of ionic reactants and the resulting ionic transition state all vary with ionic strength. Thus we have:

$$K_{\ddagger} = \frac{a_{\ddagger}}{a_{\mathrm{A}} \cdot a_{\mathrm{B}}} = \frac{[\ddagger]}{[\mathrm{A}][\mathrm{B}]} \cdot \frac{\gamma_{\ddagger}}{\gamma_{\mathrm{A}} \cdot \gamma_{\mathrm{B}}} \tag{5.17}$$

where a_X and γ_X denote the activity and the activity coefficient of species X. If we do not ignore the deviation of the last factor of this equation from unity then we obtain, in place of equation (4.22):

$$k_2 = \frac{k_{\ddagger}[\ddagger]}{[\mathrm{A}][\mathrm{B}]} = \frac{k_{\mathrm{B}}T}{h} \cdot K_{\ddagger} \cdot \frac{\gamma_{\mathrm{A}}\gamma_{\mathrm{B}}}{\gamma_{\ddagger}} \tag{5.18}$$

For aqueous solutions at 298 K, the extended Debye–Hückel theory leads, for ion i of charge $z_i e$, to the equation,

$$\log_{10} \gamma_i = -0.51 z_i^2 \frac{\sqrt{I}}{1 + \sqrt{I}} \tag{5.19}$$

where I denotes the ionic strength of the solution in the units mol dm^{-3}.

In substituting from equation (5.19) into equation (5.18), after taking logarithms of both sides of the latter, it is assumed that the activity coefficient of the transition state is obtainable from equation (5.19) by putting $z_{\ddagger} = (z_{\mathrm{A}} + z_{\mathrm{B}})$. Thus we obtain,

$$\log_{10} k_2 = \log_{10}\left(\frac{k_{\mathrm{B}}TK_{\ddagger}}{h}\right) - 0.51 \frac{\sqrt{I}}{1 + \sqrt{I}} \left[z_{\mathrm{A}}^2 + z_{\mathrm{B}}^2 - (z_{\mathrm{A}} + z_{\mathrm{B}})^2\right]$$

$$= \log_{10} k_2^{\mathrm{o}} + 1.02 z_{\mathrm{A}} z_{\mathrm{B}} \frac{\sqrt{I}}{1 + \sqrt{I}} \tag{5.20}$$

where k_2^{o} denotes the hypothetical value of the second order rate constant when all three activity coefficients are unity.

Examples of the application of equation (5.20) are shown in Figure 5.3, where the value of k_2^{o} may be deduced by extrapolating the results to zero ionic strength. These plots illustrate that the gradient is a measure of the product of the charges on the ions.

The basic theory which leads to equation (5.20) was put forward by Brønsted in 1922 and was completed in 1924 when the Debye–Hückel expression for the activity coefficient was available. Since the essence of the argument involved is so necessarily that of the Transition State Theory, it could be argued that the fundamentals of TST date from a decade

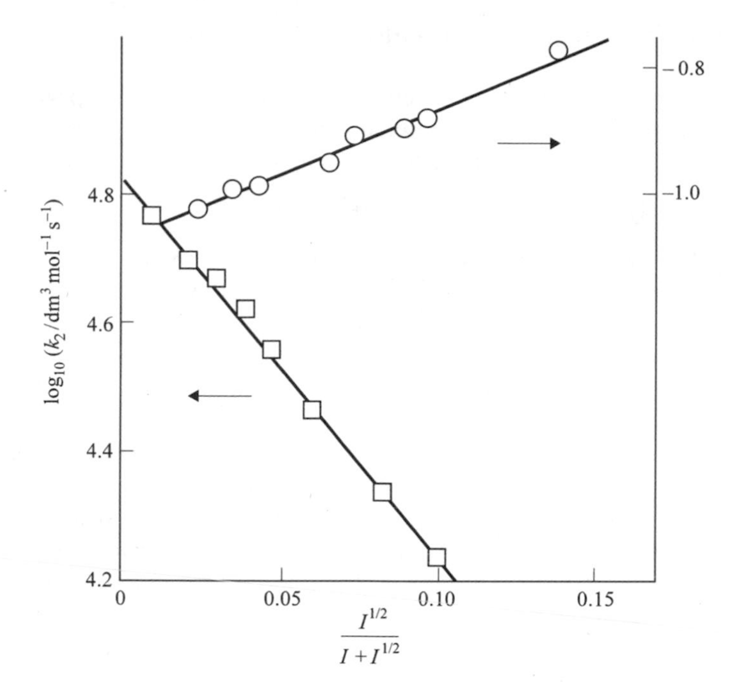

Fig. 5.3 Plots of $\log_{10} k_2$ against $I^{1/2}/(1 + I^{1/2})$ for the reactions, studied in aqueous solution: (a) $I^- + S_2O_8^{2-}$, $z_A z_B = +2$ (○, r.h. scale); (b) $Fe^{2+} + Co(C_2O_4)_3^{3-}$, $z_A z_B = -6$ (□, l.h. scale). (Data from (a) Knudsen, O. M. and C. V. King, *J. Am. Chem. Soc.*, 1939, **61,** 1205–10; (b) Barrett, J. and J. H. Baxendale, *Trans Faraday Soc.*, 1956, **52**, 210–14.)

earlier than was mentioned in Chapter 4. Equation (5.20) is widely known as the Brønsted–Bjerrum equation.

In deriving equation (5.20), the application of equation (5.19) to describe the activity coefficient of the transition state has one very interesting implication. The Debye–Hückel theory describes the equilibrium distribution of ionic species in solution, which is not established instantaneously but has a finite relaxation time that varies inversely with the ionic strength. It might then appear that equation (5.20) would be applicable to the transition state only if the lifetime of the latter is appreciably longer than the ionic relaxation time. For this condition to be generally satisfied, even in dilute solutions, would require that the lifetime of the transition state should be at least 10^{-7}s.

An alternative approach is to recognise that whenever an ion in aqueous solution undergoes random diffusive motion, it is accompanied by its own ion atmosphere. Thus as two ions approach to react together, the superposition of their ionic atmospheres necessarily provides the incipient transition state with its own equilibrium ionic environment, so that the actual lifetime of the transition state is immaterial.

5.5 Linear free energy relationships

There are many instances of a reaction, such as a hydrolysis process, in which one of the reactants may be changed slightly by adding or altering a substituent group. For example, for the reaction,

$$ArCO_2CH_3 + OH^- \rightarrow ArCO_2^- + CH_3OH \tag{5.21}$$

it is possible to put various substituents, electron-donating or electron-withdrawing, on different positions on the aromatic ring without changing the basic reaction process. In seeking an explanation for the resulting variations in the rate constant, measured under defined conditions, it is logical to consider these as arising from the effect of the substituent on the Gibbs energy of activation, $\Delta G^{\ominus}_{\ddagger} = G^{\ominus}_{\ddagger} - G^{\ominus}_{A+B}$, whose value is modified by the identity and the position of the substituent. Of course, the range of substituents referred to above may be used not merely in connection with reaction (5.21), but in various other processes. In general, equilibrium measurements can be made more rapidly and with greater accuracy than can kinetic measurements, so it is convenient to take, as the standard for assessing the effect of the various substituents, the dissociation constant for the process:

$$X\text{-}C_6H_4\text{-}CO_2H \rightleftharpoons X\text{-}C_6H_4\text{-}CO_2^- + H^+ \tag{5.22}$$

If the effect of a substituent X in modifying the standard Gibbs energy change in reaction (5.22) should be *linearly* related to this effect on the Gibbs energy of activation of reaction (5.21), then a general correlation is to be expected between the pK_a of the acid dissociation and the rate constant of the hydrolysis reaction. To describe this, the formulation introduced by Hammett in 1935 is still used.

We define the substituent parameter σ_X of a substituent X as:

$$\begin{aligned} \sigma_X &= \log_{10} K_a^X - \log_{10} K_a^o \\ &= pK_a^o - pK_a^X \end{aligned} \tag{5.23}$$

where K_a^o denotes the dissociation constant for benzoic acid and K_a^X the value when substituent X is present. The parameter σ will thus have a positive value for a substituent which withdraws electrons from the aromatic ring and a negative value for one which donates electron density to the ring.

The Hammett equation is

$$\log_{10}(k^X/k^o) = \sigma_X.\rho \tag{5.24}$$

where ρ is called the reaction parameter. To test it, one may plot $\log_{10}k^X$ against σ_X for a wide range of substituents X, and if a single value of ρ is operable then this reaction parameter is the gradient of the linear plot so obtained. An illustration of this phenomenon is shown in Figure 5.4, where $\rho = +2.66$.

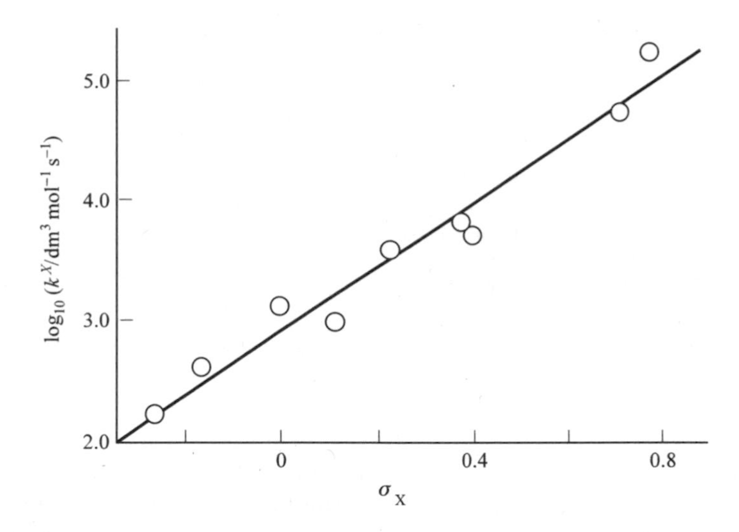

Fig. 5.4 Plot of $\log_{10}k^{\mathrm{X}}$ against the substituent parameter, σ_{X}, for the reaction of *C*-aryl,*N*-*p*-toluenesulphonylmethyleneimine with $NaBH_4$ in 2-propanol at 298 K. (Data from Henderson, C. M., 1991, DPhil thesis, University of Ulster, p. 222.)

The significance of the value of the reaction parameter ρ is perhaps more readily seen if we combine equations (5.23) and (5.24) and obtain:

$$\rho = \frac{\log_{10}(k^{\mathrm{X}}/k^{\mathrm{o}})}{\log_{10}(K_{\mathrm{a}}^{\mathrm{X}}/K_{\mathrm{a}}^{\mathrm{o}})} \tag{5.25}$$

Thus ρ is the ratio of the extent to which substituent X decreases the Gibbs energy of activation of reaction (5.21) to the extent to which it decreases the standard Gibbs energy change for process (5.22). In this case, the effect on the former is $+2.66$ times the latter, but for other reactions the ratio may be quite different and in some instances is negative. Some typical values of the reaction parameter ρ are shown in Table 5.1.

However, the Hammett relation is not in fact universally observed, despite its manifest success in cases such as that illustrated in Figure 5.4, where the changes in the substituent causes a variation in the rate constant, under defined conditions, of three orders of magnitude. Possible reasons for the occurrence of deviations include the following:

(i) Steric effects of the substituent, such that these are appreciably more important in the reaction of the substituted compound than in the dissociation of the substituted benzoic acid.

For this reason, all *ortho*-substituents are normally excluded from consideration. For example, the substituents shown in Figure 5.4. are all in the *meta*- or the *para*- positions.

Table 5.1 Typical values of the Hammett reaction parameter, ρ.

Reaction	Solvent	ρ
$ArCO_2Et + OH^-$	ethanol/water	+2.55
$ArCOCl + EtOH$	ethanol	+1.735
$ArCONH_2 + OH^-$	ethanol/water	+1.40
$ArCO_2H + Ph_2CN_2$	ethanol	+0.98
$ArCONH_2 + H^+$	ethanol/water	−0.50
$ArCO_2H + MeOH$	pyridine	−0.585
$ArNH_2 + HCO_2H$	pyridine	−1.30

Data taken from Wells, P. R., 1963, Linear free energy relationships, *Chemical Reviews*, **63**, 171–219.

(ii) The change in $G^{\ominus}_{\ddagger}$ is linear with that in $\Delta G^{\ominus}$ over part but not all of the substituent range, leading to a non-linear Hammett plot. Possible reasons for this may be deduced from a detailed consideration of the reaction mechanism.

(iii) If the rate constant is initially quite large, the presence of substituents which bring about an acceleration of the rate may require, for a linear Hammett plot, rate constants far in excess of the diffusion-controlled limit. Sometimes this particular problem may be avoided by the choice of the conditions of solvent and temperature under which the rate constant is to be measured.

It should also be admitted that while the original basis for determining substituent constants was as detailed in equation (5.23), this procedure was implicitly abandoned when it was found that a modified value of σ_X would lead to a more linear Hammett plot. There have also been refinements with regard to the type of reaction process to which a particular set of σ values apply.

While the Hammett equation is restricted to aromatic systems, a comparable correlation known as the Taft equation,

$$\log_{10}(k^X/k^o) = \sigma_X^* \cdot \rho^* \tag{5.26}$$

has been introduced for aliphatic systems. Though not strictly analogous in the sense that the values of σ_X^*, the substituent parameter, are not based on differences in the dissociation constants of certain acids, it is, like the Hammett equation, based on the premise of linearly related changes in the Gibbs energy.

5.6 Kinetic isotope effects

As explained in the previous chapter, one of the chief obstacles to the use of Transition State Theory to produce an absolute value for a rate constant is that there are extreme difficulties in the exact evaluation of potential energy surfaces. However, when the aim is simplified to obtaining an understanding of the changes in the rate constant consequent upon isotopic substitution, this particular problem vanishes since it can be implicitly assumed that the potential energy surface, whatever its shape, will be unaffected by a slight change in the mass of one nucleus.

If the atom concerned in the isotopic replacement is one which is not directly involved in the bond breaking or bond forming that occurs during the course of the reaction, then the effect is described as *secondary* kinetic isotope effect. Where the atom in question is directly involved in the reaction, then we have a *primary* kinetic isotope effect. In general the magnitude of the effect for the latter group is much more substantial than for the former.

Since the size of a kinetic isotope effect (KIE) depends on the ratio of, rather than on the difference in, the masses of the two isotopes concerned, the effect of substituting a deuterium atom for a hydrogen atom is greater than that of any other isotopic substitution, with the exception of a tritium atom for a hydrogen atom. It is convenient to represent the magnitude of a kinetic isotope effect, arising from the substitution of D for H, as the ratio of the rate constant for the protium-containing species to that of the deuterium-containing analogue, often denoted as k_H/k_D.

In making a cursory analysis of the magnitude of the effect, it is convenient to start from equations (4.22) and (4.23) in the previous chapter.

$$k_2 = \kappa \frac{k_B T}{h} \cdot \frac{q_{\ddagger}^{0}}{q_A^{0} \cdot q_B^{0}} \, e^{-\Delta\varepsilon^{o}/k_B T} \tag{5.27}$$

Let us further assume:

(i) that the kinetic isotope effect arises from the substitution of D for H in reactant A,
(ii) that this substitution does not alter the symmetry number of species A or of the transition state, and
(iii) that in regard to vibrational energy, $\varepsilon = 0$ represents the minimum of each potential energy curve, so that each vibrational partition function is of the form,

$$q_{vib} = \frac{e^{-h\nu/2k_B T}}{1 - e^{-h\nu/k_B T}} \tag{5.28}$$

where ν represents the vibration frequency. It follows, since $\Delta\varepsilon^{o}$ is precisely the same for each reaction, that the kinetic isotope effect is given by:

$$\frac{k_H}{k_D} = \frac{q_{\ddagger}^{0}(H) \cdot q_A^{0}(D)}{q_{\ddagger}^{0}(D) \cdot q_A^{0}(H)} \tag{5.29}$$

where $q_A^0(D)$ and $q_A^0(H)$ distinguish the reduced partition functions of reactant A when deuterium is present from that when protium is present, and a similar distinction is made in regard to the partition functions of the deuterated and undeuterated transition states.

Each molecular partition function in equation (5.29) is the product of factors representing translation, rotation and vibration. It can readily be shown that the quotient of the translational partition functions in equation (5.29) is very close to unity, since the higher mass of deuterated A in the numerator is largely counteracted by the higher mass of the deuterated transition state in the denominator. The same applies to the rotational partition functions. With regard to the vibrational partition functions, in many cases the vibration frequency of A or of the transition state is unaffected by isotopic substitution so that these factors cancel out. In the case of a primary kinetic isotope effect, we are left, typically, with three vibrations of reactant A—one stretching and two bending—and two of the transition state, both bending, whose frequencies are substantially affected by substituting D for H.

Since these vibrations are all of moderate or high frequency, in each case $h\nu \gg k_B T$ and so it is justified to replace equation (5.28) by $q_{vib} \approx e^{-h\nu/2k_B T}$. So the predominant consequence of isotopic substitution is the concomitant change in the zero-point energy (ZPE) of every vibration involving the atom being substituted.

In the case of a primary kinetic isotope effect, the stretching vibration in the reacting molecule A, involving H or D, will have no counterpart in the transition state. The magnitude of the kinetic isotope effect depends considerably on how the bending vibrations change when the transition state is formed. There are three main possibilities here and it is instructive to look at the consequences of each.

(i) Bending frequencies unchanged. If the bending vibrations involving the H (or D) are the same in the transition state as in the reactant molecule, then the corresponding vibrational partition functions cancel out, leaving the difference in zero-point energies of the stretching vibration in the unsubstituted and substituted reactant molecules as the single cause of the rate change on isotopic substitution. This situation is illustrated diagrammatically in Figure 5.5, and leads to:

$$\frac{k_H}{k_D} = \frac{e^{-h\nu_A(D)/2k_B T}}{e^{-h\nu_A(H)/2k_B T}} = \exp h[\{\nu_A(H) - \nu_A(D)\}/2k_B T] \tag{5.30}$$

(ii) Bending frequencies decreased in the transition state. If the bending frequencies should not be the same in the transition state as in the reactant molecule, then there are three vibrational frequencies of the reactant molecule which are altered by isotopic substitution and which do not cancel out between the reactant and the corresponding transition state. Thus we have:

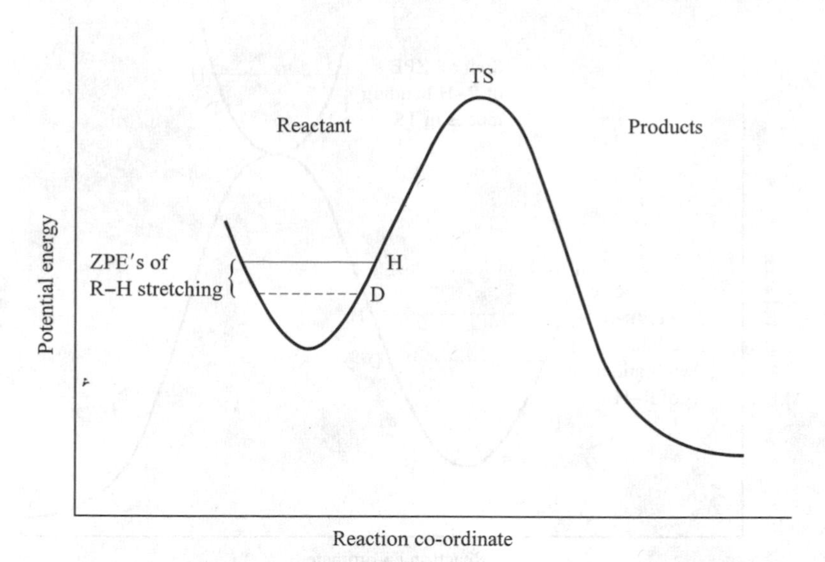

Fig. 5.5 Illustration of the effect of deuterium substitution on the rate constant when the bond being broken has identical bending frequencies in the reactant and the transition states (TS).

$$\frac{k_{\mathrm{H}}}{k_{\mathrm{D}}} = \exp\left\{\frac{h}{2k_{\mathrm{B}}T}\left[\sum^{3}\{\nu_{\mathrm{A}}(\mathrm{H}) - \nu_{\mathrm{A}}(\mathrm{D})\} - \sum^{3}\{\nu_{\ddagger}(\mathrm{H}) - \nu_{\ddagger}(\mathrm{D})\}\right]\right\} \tag{5.31}$$

For this circumstance a more intricate diagram is necessary, as shown in Figure 5.6. Where the bending frequencies in the transition state are less than in reactant A, then the kinetic isotope effect should be enhanced beyond the value in (i).

(iii) Bending frequencies increased in the transition state. Were this to come about, then a modification of Figure 5.6 would be required to represent it. The value of the kinetic isotope effect would now be less than in case (i), to an extent depending on the magnitude of this increase in frequency.

Given the form of equations (5.30) and (5.31), it is clear that $k_{\mathrm{H}}/k_{\mathrm{D}}$ is temperature dependent and decreases as T is increased. Thus it is important when comparison is to be made between two such values—whether one is experimental and one calculated or both are experimental—that both refer to the same temperature.

To make an estimate of a typical kinetic isotope effect, let us assume that the vibration frequencies of the R–H bond in reactant A are at 3100 cm^{-1} (stretching) and 1100 cm^{-1} (bending). Since the H/D atom is very light in

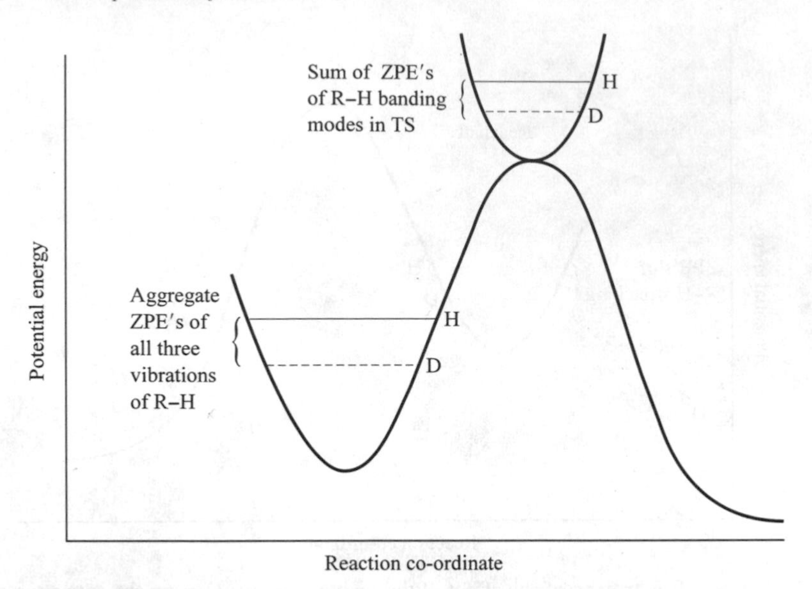

Fig. 5.6 Illustration of the effect of deuterium substitution on the rate constant when the bond being broken has different bending frequencies in the reactant and the transition states.

relation to the rest of the molecule, the reduced mass of R–D may be estimated as twice that value. The vibration frequency is inversely proportional to the square root of the reduced mass so this leads, as a rough approximation, to 2191 cm^{-1} (stretching) and 778 cm^{-1} (bending) as the corresponding "frequencies" of the deuterated species. Thus we have, for the case where the bending frequencies are unchanged in the transition state,

$$\frac{k_H}{k_D} = e^{0.5hc(3100-2192)/298k_B} = 9.0 \tag{5.32}$$

In a study of the dissociation of gaseous $CHCl_3$ and of $CDCl_3$ under electron impact, it was found that, at 518 K, the former lost an H atom 3.0 times more readily than the latter lost D. When the ratio in equation (5.32) is re-evaluated at the temperature of this experiment it gives k_H/k_D = 3.5. Using the actual (known) frequencies of $CHCl_3$ and $CDCl_3$ brings the estimate closer to the experimental measurement.

Many secondary kinetic isotope effects are so small that k_H/k_D differs from unity by only a few percent, but instances of much higher values are known. One example comes from the solvolysis, in aqueous ethanol, of *cis*-4-*tert*-butylcyclohexyl brosylates, where the size of the *tert*-butyl group forces the leaving group to an axial position. Replacement of H by D in specified positions produced secondary kinetic isotope effect values as shown in Table 5.2.

Table 5.2 Secondary kinetic isotope effects in the solvolysis of *cis*-4-*t*-butylcyclohexylbrosylate in aqueous ethanol at 308 K.

Site of deuteration	k_H/k_D
α, equatorial	1.20
β, axial	1.44
β, equatorial	1.10

Data taken from Shiner, V. J. Jr and J. G. Jewett, 1965, *J. Am. Chem. Soc.*, **87**, 1382–3.

These results show that the presence of a D atom in the β-axial position has an appreciable effect, indicating a significant reduction in the aggregate of the vibration frequencies of the β-axial hydrogen on forming the transition state. The size of this change, in contrast to that regarding the β-equatorial hydrogen, is consistent with a specific involvement of this atom in the solvolysis process.

5.7 Electron transfer reactions in solution

The most familiar of electron transfer reactions in solution are those involving the ions of transition metals, such as:

$$Ce^{4+} + Fe^{2+} \rightarrow Ce^{3+} + Fe^{3+} \tag{5.33}$$

In a reaction of this type no bonds are broken or formed, which is a clear distinction from reactions considered in Chapter 4 in regard to the development of Transition State Theory. However, in addition to the transfer of an electron, there are other changes as a consequence of the reaction, in that the equilibrium environment of solvent molecules is not the same around Fe^{3+} as around Fe^{2+}, nor around Ce^{3+} as around Ce^{4+}.

It is possible to make qualitative assessments of the rates of some electron transfer reactions by considering the concomitant changes in the solvent environment. Self-exchange reactions, where the products are chemically the same as the reactants, may be studied by using radio-isotopes as labels

$$Ru^{2+} + {}^*Ru^{3+} \rightarrow Ru^{3+} + {}^*Ru^{2+} \tag{5.34}$$

For such reactions, where the standard Gibbs energy change is zero, the forward and reverse rate constants are equal. However, the rate constants for self-exchange show considerable variation.

In a self-exchange reaction, the extent of the rearrangement of the solvent environment depends on the size of the ion and on the change in ionic size which accompanies electron transfer. For reaction (5.34), appreciable solvent reorganisation is required and consequently it is expected to be slow. On the other hand, for the reaction,

$$Ru(bpy)_3{}^{2+} + {}^*Ru(bpy)_3{}^{3+} \rightarrow Ru(bpy)_3{}^{3+} + {}^*Ru(bpy)_3{}^{2+} \quad (5.35)$$
$$(\text{bpy} \equiv 1,\ 1\text{-bipyridyl})$$

where the ions are considerably larger, a much faster exchange reaction is to be anticipated. Measured rate constants are around $60\,dm^3\,mol^{-1}\,s^{-1}$ for reaction (5.34) and around $4 \times 10^8\,dm^3\,mol^{-1}\,s^{-1}$ for reaction (5.35).

A more quantitative approach to this question has been developed by Marcus, the 1992 Nobel Laureate, who recognised that the reaction pathway should be one in which energy is conserved, and thus that it is not realistic to envisage the electron transfer as occurring first, followed by all the requisite solvent reorganisation.

Around an ion in solution there will be an equilibrium or most probable configuration of solvent molecules, which represents the Gibbs energy minimum. Momentary fluctuations from this configuration will occur, as the position and the orientation of individual solvent molecules is altered. These, whether they involve the solvent molecule coming closer to the ion or going farther from it, must be identified with higher values of the Gibbs energy. While these fluctuations in the value of G are a function of many inter-molecular distances and angles, it is conceivable that all of these parameters may be combined so that G may be expressed as a function of one aggregate solvent parameter.

On this basis, we may conceive of two curves of Gibbs energy against nuclear co-ordinates, one pertaining to the reactants plus solvent environment, the other to products plus solvent environment. These curves, shown in Figure 5.7, are deduced to be essentially parabolic, and their point of

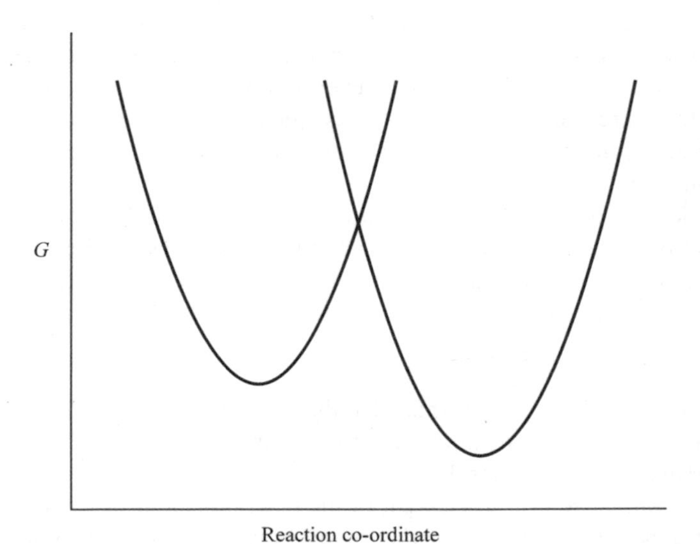

Fig. 5.7 Plots of the Gibbs energy, G, of (l.h. curve) reactants plus environment and (r.h. curve) products plus environment, both against the "reaction co-ordinate", relating to an electron transfer reaction.

intersection represents the same configuration of the solvent around reactants and products. Marcus obtained an expression for the rate constant for electron transfer, assuming that this occurred only at such a configuration where the Gibbs energies of reactants and products are equal, and simplified it to give the equation

$$k_r = A \exp\left\{\frac{-\lambda}{4RT}\left(1 + \frac{\Delta G^{\ominus}}{\lambda}\right)^2\right\} \tag{5.36}$$

where A is a factor which depends on the type of electron transfer involved, λ is a solvent reorganisation term and $\Delta G^{\ominus}$ is the standard Gibbs energy change for the reaction.

For a self-exchange reaction, such as (5.34), $\Delta G^{\ominus}$ is zero so that equation (5.36) becomes:

$$k_r = A \exp\left(\frac{-\lambda}{4RT}\right) \tag{5.36a}$$

This equation is totally consistent with the qualitative argument presented earlier regarding the rates of self-exchange reactions, in relation to the relative rates of reactions (5.34) and (5.35).

From these equations it is possible to deduce an interesting relation for the rate constant of a redox reaction in terms of those of the relevant self-exchange reactions. For example, for the reaction,

$$Co(phen)_3{}^{3+} + Ru(NH_3)_6{}^{2+} \rightarrow Co(phen)_3{}^{2+} + Ru(NH_3)_6{}^{3+} \tag{5.37}$$

if the rate constant and the equilibrium constant are denoted by k_{12} and K_{12} then it may be shown that, to a good approximation,

$$k_{12} \approx (k_{11}k_{22}K_{12}f)^{1/2} \tag{5.38}$$

where k_{11} and k_{22} denote the rate constants for self-exchange of $Co(phen)_3{}^{3+}/Co(phen)_3{}^{2+}$ and for $Ru(NH_3)_6{}^{3+}/Ru(NH_3)_6{}^{2+}$ and f is a factor which in certain cases can be shown to be close to unity. Equation (5.38) is known as the Marcus cross-relation.

The rate constant for reaction (5.37) has been measured as $1.5 \times 10^4\,dm^3\,mol^{-1}\,s^{-1}$, and those of the self-exchange reactions are known: $k_{11} = 4 \times 10^1\,dm^3\,mol^{-1}\,s^{-1}$ and $k_{22} = 3 \times 10^3\,dm^3\,mol^{-1}\,s^{-1}$. Also, the equilibrium constant for reaction (5.37), K_{12}, is known to be 2.6×10^5. If we neglect the deviation of f from unity, from equation (5.37) we may estimate k_{12}, the rate constant for reaction (5.37), as $1.8 \times 10^5\,dm^3\,mol^{-1}\,s^{-1}$. This value is high by an order of magnitude, but bearing in mind the range of rate constants found for electron transfer reactions, it is clearly a useful estimate. A marginal improvement in this first order approximation can be obtained by evaluating the factor f.

Another interesting feature of Marcus theory is the manner in which the rate constant of a redox process is predicted to vary over a series of reactions where the value of λ, the solvent reorganisation term, remains constant. Rearranging equation (5.36) and taking logarithms of both sides we have:

$$\ln k_r = \ln A - \frac{1}{4\lambda RT}(\lambda + \Delta G^{\ominus})^2 \tag{5.37a}$$

This equation shows that the rate constant should be a maximum when $\Delta G^{\ominus} = -\lambda$, and that it should decrease, both when $\Delta G^{\ominus}$ becomes less negative, which is readily understandable, but also when $\Delta G^{\ominus}$ becomes more negative. This last prediction tends to run counter to chemical intuition and for some time it seemed not to be in accord with the behaviour observed experimentally.

The physical basis of these predictions may be understood with the aid of Figure 5.8. The parabola on the left represents the reactants. The middle curve on the right intersects it at the base because $\Delta G^{\ominus}$ for this reaction has the optimum (negative) value, and so it is predicted that $\Delta G^{\ominus}_{\ddagger}$ of this reaction will be close to zero. Hence reaction (a) should be very rapid. The top curve represents a reaction for which $\Delta G^{\ominus}$ is only slightly negative and from the position of its intersection with the curve on the left, it is seen that this reaction will be rather slower. This illustrates the decline in the rate constant which occurs in the Marcus "normal" region. The bottom curve represents a reaction for which $\Delta G^{\ominus}$ is more negative than is reaction (a), but the point of intersection of the two parabolas implies that its rate of reaction will also

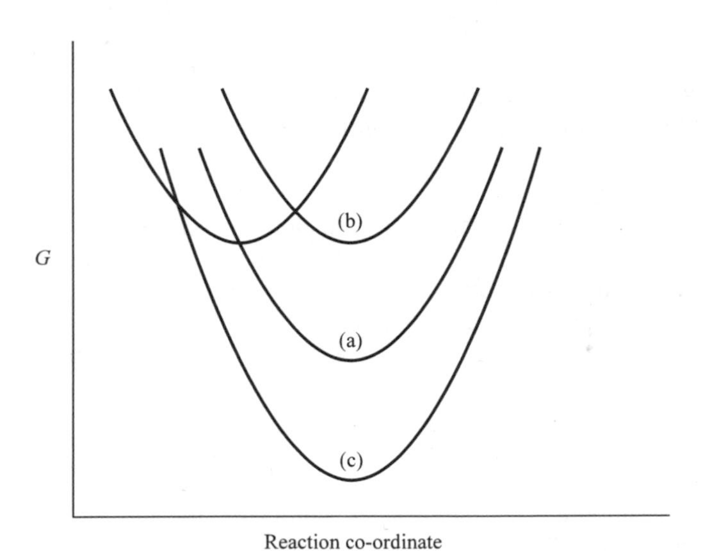

Fig. 5.8 Plots of the Gibbs energy, G, of the reactants plus solvent environment (l.h. curve) and (r.h. curves) of products plus solvent environment for three distinct cases: (a) $\Delta G^{\ominus}$ appreciably negative, barrier height ≈ 0; (b) $\Delta G^{\ominus} = 0$, finite barrier height, and (c) $\Delta G^{\ominus}$ very highly negative, but a similar finite barrier height.

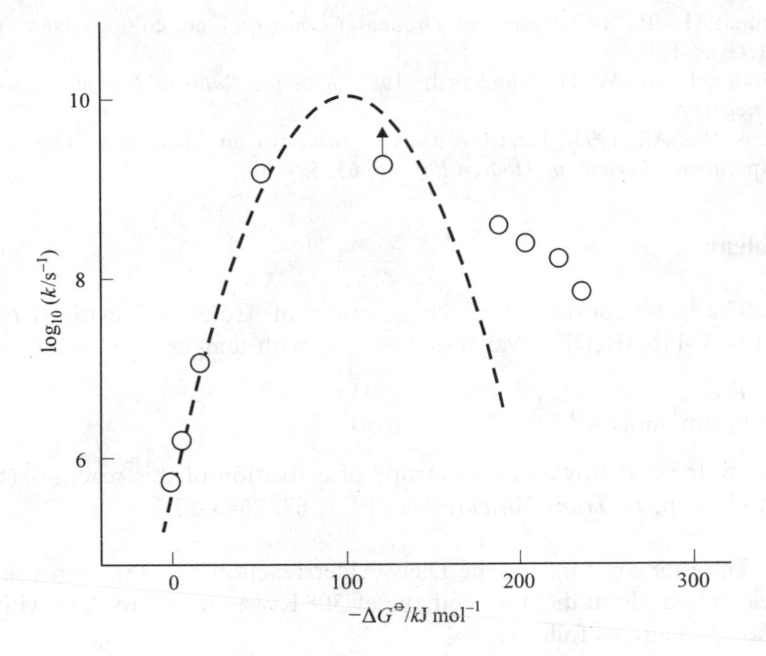

Fig. 5.9 Plot of $\log_{10}(k/s^{-1})$ for the intramolecular exchange of a series of radical ions against $-\Delta G^{\ominus}$. The dotted curve is calculated from equation (5.37a). (Data from Miller, J. R., T. Calcaterra and G. L. Closs, *J. Am. Chem. Soc.*, 1984, **106**, 3047–9.)

be slower. Curve (c) exemplifies what is known as the Marcus "inverted" region.

Attempts to test equation (5.36b) experimentally have reproduced the expected increase in the rate constant as $\Delta G^{\ominus}$ becomes slightly negative, but in some systems, as $\Delta G^{\ominus}$ becomes increasingly negative, the rate constant attains and remains at a plateau value, rather than showing the decline predicted by equation (5.37b) in the Marcus inverted region. An example of this type of behaviour is shown in Chapter 11 in regard to the quenching of fluorescence. In 1984, in studying intramolecular electron transfer between the two ends of a molecule of which the centre was a "spacer" of constant size, Miller, Calcaterra and Closs found that the first order rate constant decreased appreciably at higher values of $-\Delta G^{\ominus}$, much as predicted by Marcus theory, as is illustrated in Figure 5.9.

Suggested reading

Kohnstam, G., 1970. The kinetic effect of pressure. *Progress in Reaction Kinetics*, **5**, 335–408.

Perlmutter-Hayman, B., 1971. The primary kinetic salt effect in aqueous solution. *Progress in Reaction Kinetics*, **6**, 239–67.

Hammett, L. P., 1970. *Physical Organic Chemistry*, 2nd edition. New York: McGraw-Hill.

Melander, L. and W. H. Saunders Jr, 1987. *Reaction Rates of Isotopic Molecules.* Krieger.

Marcus, R. A., 1993. Electron transfer reactions in chemistry. Theory and experiment. *Reviews of Modern Physics*, **65**, 599–610.

Problems

5.1 The rate constant for the reaction of $Co(en)_3{}^{3+}$ with $[Cr(2,2'\text{-}bipy)_2(4,4'\text{-}bipy)H_2O]^{2+}$ was found to vary with temperature as follows:

T/K	288.1	304.8
$k_2/\mathrm{dm^3\,mol^{-1}\,s^{-1}}$	718	1050

Evaluate the enthalpy and the entropy of activation of this reaction. (Data from Ulstrup, J., *Trans Faraday Soc*, 1971, **67**, 2645–51.)

5.2 The rate constant for the Diels–Alder reaction between isoprene and maleic anhydride in dichloromethane at 308 K was found to vary with the applied pressure as follows:

p/atm	1	341	681	1022
$k/10^{-4}\,\mathrm{s^{-1}}$	5.28	8.25	12.6	18.9

(1 atm $\equiv 1.013 \times 10^5\,\mathrm{N\,m^{-2}}$)

Evaluate the mean activation volume of the reaction over this pressure range. (Data from Grieger, R. A. and C. A. Eckert, *Trans Faraday Soc.*, 1970, **66**, 2579–84.) (N.B. These authors used mole fractions rather than molarities as units of concentration, hence the units of the rate constant.)

5.3 In a study of the alkaline hydrolysis of methyl acetate, the rate constant was found to vary with temperature as follows:

T/K	293	298	303	308
$k/\mathrm{dm^3\,mol^{-1}\,s^{-1}}$	0.135	0.186	0.259	0.358

Hence evaluate the standard enthalpy of activation and the entropy of activation for this reaction. (Data from Saadi, A. H., and W. H. Lee, *J. Chem. Soc., B*, 1966, 1–4.)

5.4 For the reaction of persulphate with ferrocyanide ions, in aqueous solution at 298 K, added $KClO_4$ was found to increase the rate constant as follows:

$I/\mathrm{mol\,dm^{-3}}$	0.0599	0.0663	0.0696	0.0728	0.0792
$k/\mathrm{dm^3\,mol^{-1}\,min^{-1}}$	0.55	0.64	0.70	0.74	0.86

How well do these data fit the Brønsted–Bjerrum equation? (Data from Kershaw, M. R. and J. E. Prue, *Trans Faraday Soc.*, 1967, **63**, 1198–1207.)

5.5 From competition studies in aqueous solution at 298 K in which the ionic strength, I, was varied, the relative rate constant, $k(\mathrm{A + B})/k(\mathrm{A + C})$, was found to depend on ionic strength as follows:

I/mol dm^{-3}	0.01	0.16
$k(\mathrm{A + B})/k(\mathrm{A + C})$	4.82	3.11

(a) Assuming that the reactions of A with B and with C obey the Brønsted–Bjerrum equation, what deductions may be made about the species A, B, C ?

(b) Given that B is H_2O_2 and C is NO_2^- ion, what is the charge on A?

5.6 In the reduction of *C*-aryl,*N*-benzenesulfonylimines by lithium tri (*tert*-butoxy) aluminium hydride,

$$X\text{-}C_6H_4\text{-}CH{=}N\text{-}SO_2\text{-}C_6H_5 + LiAl(O\,Bu^t)_3H$$

the rate constant was found to vary with the aryl substituent X as follows:

X	$k/10^3\ \mathrm{dm^3\ mol^{-1}\ s^{-1}}$	σ_X^+
H	2.2	0
p-CH_3	0.95	−0.31
m-Cl	6.7	+0.37
p-NO_2	16.3	+0.79
p-OCH_3	0.289	−0.78

Thus determine the reaction constant ρ for the reduction process involved. (Data from Hanly, A. M., DPhil thesis, University of Ulster, 1994.)

5.7 For the reaction of $CF_3^{\cdot}$ radical with $NH_3(ND_3)$ to yield $CF_3H(CF_3D)$, it is reported that at 423 K, $k_H/k_D = 4.3$. Is this consistent with the zero point energy difference between the stretching vibrations of N–H and N–D in ammonia being 5.4 kJ mol^{-1}? (Data from Gray, P., N. L. Arthur and A. C. Lloyd, *Trans Faraday Soc.*, 1969, **65**, 775–9.)

5.8 In solvolysis studies of *cis*-4-*tert*-butylcyclohexylbrosylate, in 50% aqueous ethanol at 308 K, single and multiple deuteration at known sites was found to alter the rate constant as follows:

Deuteration	$k_1/10^{-4}\ \mathrm{s^{-1}}$
d_0	3.938
$\beta(a)$, d_1	2.743
$\beta(e)$, d_1	3.593
$\beta(a, a, e)$, d_3	1.535
$\beta(a, a, e, e)$, d_4	1.376

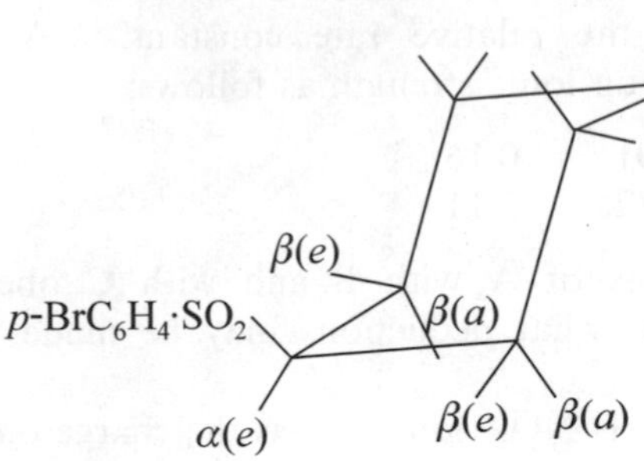

^{t}Bu

$\beta(e)$

p-$BrC_6H_4 \cdot SO_2$ $\beta(a)$

$\alpha(e)$ $\beta(e)$ $\beta(a)$

Show that these data provide evidence that the transition state for the solvolysis reaction does not have a plane of symmetry. (Data from Shiner, V. J. Jr, and J. G. Jewett, *J. Am. Chem. Soc.*, 1965, **87**, 1382–3.)

5.9 For the electron transfer reaction,

$$Ru(NH_3)_6^{2+} + Fe^{3+} \rightarrow Ru(NH_3)_6^{3+} + Fe^{2+}$$

the equilibrium constant $K_{12} = 5.0 \times 10^{11}$. The rate constants of self-exchange of the reactants are as follows:

$Ru(NH_3)_6^{3+/2+}$	$k = 4.3 \times 10^3\ dm^3\ mol^{-1}\ s^{-1}$
$Fe^{3+/2+}$	$k = 4\ dm^3\ mol^{-1}\ s^{-1}$

Using the Marcus expression and taking f as unity, estimate the rate constant for the above electron-transfer reaction. (The experimental value is $3.4 \times 10^5\ dm^3\ mol^{-1}\ s^{-1}$.) (Data from Chou, M., C. Creutz and N. Sutin, *J. Am. Chem. Soc.*, 1977, **99**, 5615–23, and Weaver, M. J. and E. L. Yee, *Inorg. Chem.*, 1980, **19**, 1936–45.)

6 Unimolecular gas phase reactions

Several gas phase reactions are known which follow first order kinetics, but with the unusual feature that the value of the first order rate constant, measured at constant temperature, depends on the gas pressure. These reactions are accepted as unimolecular processes, but the variations in the first order rate constant have acted as a spur to the theoretical explanation of reactions showing such an idiosyncrasy.

6.1 The kinetic results

For many decades, a number of decomposition or rearrangement reactions in the gas phase have been known in which the concentration of the reacting species decays exponentially with time, following first order kinetics. The simplest mechanism consistent with this behaviour is, of course, a unimolecular process and this was proposed as the mechanism involved.

Further study demonstrated that in two respects the topic was more complex than was at first apparent. Firstly, some reactions, initially thought to be in this category, were shown not to take place by an elementary process, but rather to occur by a chain reaction involving free radicals. Secondly, it was shown that for the other more authentic examples, while the reaction in a closed volume at constant temperature always followed first order kinetics, the first order rate constant was dependent on the gas pressure. It was not at all clear that such behaviour could be consistent with the proposed unimolecular mechanism.

Examples of such reactions are the isomerisation of cyclopropane to propene, or of *cis*-1,2-dimethylcyclopropane to the *trans* isomer. In some cases, there may be two product molecules: for example, iodoethane yields hydrogen iodide and ethene, while cyclobutane dissociates to give two molecules of ethene and methylcyclobutane gives propene plus ethene. In some instances, the unimolecular dissociation may yield a free radical species. For example, ethane dissociates to two methyl radicals and azomethane yields N_2 plus two methyl radicals. In studying any of these it is important to avoid kinetic complications arising from subsequent reactions of any product of the unimolecular reaction under study.

An illustration of the nature of these variations in the first order rate constant, k', is shown in Figure 6.1. Above a certain pressure, k' is constant

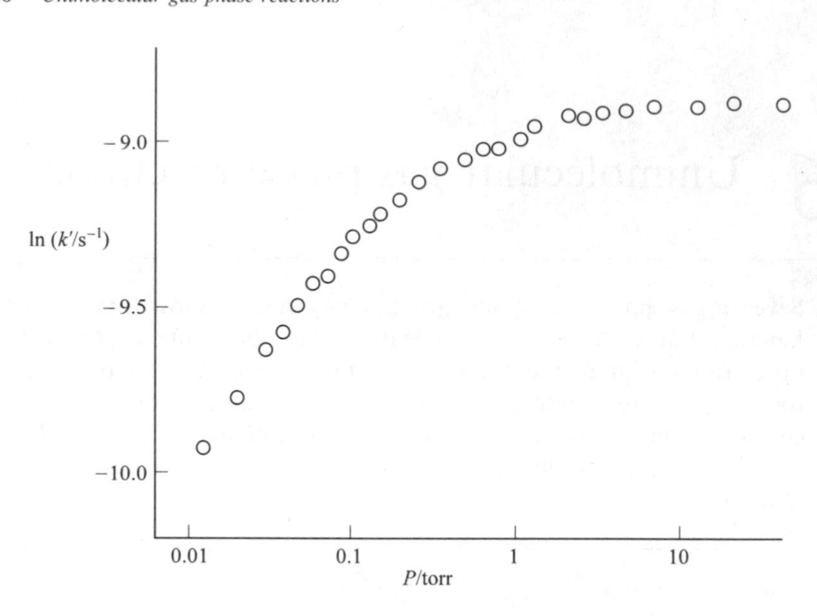

Fig. 6.1 Plot showing the variation of the first order rate constant, k', with the pressure, P, for the isomerisation of 3-methylcyclobutene at 396.6 K. (Data from Frey, H. M. and D. C. Marshall, *Trans Faraday Soc.*, 1965, **61**, 1715–21.)

but as the pressure is decreased, k' shows a falling away from this high pressure limit. This decrease becomes progressively steeper at lower pressures until, in the limit, k' becomes proportional to the pressure. However, this transition from the constant value of k' at the high pressure limit to the stage where the plot of k' is pointing towards the origin is a very gradual one, extending over a pressure range of perhaps four orders of magnitude.

The effect of adding a chemically stable gas was also investigated. Where the pressure of the reactant gas is above the high pressure limit, the addition of a gas such as argon or nitrogen has no effect on the reaction. But at a pressure below the limit at which the "plateau" value of k' was observed, the addition of an unreactive gas causes an increase in k', eventually raising it up to the value at the high pressure limit. While equal pressures of different gases may cause k' to rise by different extents, the limiting values of k' at high inert gas pressures are all the same and equal to that achieved at a high pressure of the pure reactant.

Another unusual feature of unimolecular gas reactions is the nature of their temperature dependence. When experiments are conducted over a range of temperatures, it is found that the activation energy is constant at pressures above the high pressure limit, but at lower pressures it progressively decreases. For example, for the isomerisation of cyclopropane to propene, above a pressure of 200 torr, E_a is 274.5 kJ mol^{-1}, but at 5 torr it has fallen to 254.3 kJ mol^{-1}. This decrease is by an amount appreciably greater than the limits of experimental error and thus deserves an explanation.

6.2 The Lindemann mechanism

The first step in explaining the variability of the first order rate constant of a unimolecular gas reaction is frequently attributed to Lindemann, but there is a rival claim from Christiansen on the basis of his doctoral thesis at Copenhagen.

The significant ideas put forward in 1922 were (i) that reactant molecules might need to acquire much higher than average vibrational energy in order for reaction to be possible, and that this could occur in the course of intermolecular collisions, and (ii) that a molecule with excess vibrational energy, capable of reacting, might have a finite lifetime before it undergoes reaction. Accepting that the mean interval between inter-molecular collisions is inversely proportional to the gas pressure, then above a certain pressure, collisions should be sufficiently frequent that the equilibrium concentration of highly energised molecules is preserved. However, at much lower pressures, there will be such a long interval between collisions that most highly energised molecules will have reacted long before the next collision occurs. Consequently, the concentration of energised molecules would be significantly less than the equilibrium value, so that at these low pressures the rate constant would be less than at high pressures.

The Lindemann theory is frequently summarised in terms of the following mechanistic steps, using the nomenclature that A represents a molecule of the isomerising substance, and A^* a reactant molecule that has acquired sufficient excess vibrational energy to undergo reaction to yield the product B:

$$\left.\begin{array}{rcl} A + A & \underset{-1}{\overset{1}{\rightleftarrows}} & A^* + A \\ A^* & \overset{2}{\rightarrow} & B \end{array}\right\} \tag{6.1}$$

While this mechanism is correct in the initial phases of the reaction, it is seriously defective at every other stage. Its use to describe the progress of the reaction has led to much unnecessary confusion.

The significant point overlooked in mechanism (6.1) is that a reactant molecule A may become energised by collision with any other molecule. In the present system one must then add the effect of its collision with a product molecule, B:

$$A + B \underset{-3}{\overset{3}{\rightleftarrows}} A^* + B \tag{6.2}$$

Applying the Principle of Stationary States to the species A^*, we now have:

$$\frac{d[A^*]}{dt} = k_1[A]^2 + k_3[A][B] - k_{-1}[A][A^*] - k_{-3}[B][A^*] - k_2[A^*] = 0 \tag{6.3}$$

Thus

$$[A^*] = \frac{k_1[A]^2 + k_3[A][B]}{k_{-1}[A] + k_{-3}[B] + k_2} \tag{6.4}$$

It is convenient to use $(a - x)$ for the concentration of A at time t and x for that of B:

$$[A^*] = \frac{\{k_1(a-x) + k_3x\}(a-x)}{k_{-1}(a-x) + k_{-3}x + k_2} \tag{6.4a}$$

Thus the rate of step *2*, in which the product is formed, is given by:

$$\frac{dx}{dt} = k_2[A^*] = \frac{k_2\{k_1(a-x) + k_3x\}(a-x)}{k_{-1}(a-x) + k_{-3}x + k_2} \tag{6.5}$$

Since B, the reaction product, is an isomer of A, these molecules must have the same molecular weight and be of similar size. It may then be expected that their rate constants for energising A and for de-energising A^* will be fairly similar, i.e., that $k_1 \approx k_3$ and $k_{-1} \approx k_{-3}$. If these are treated as strict equalities*, we have:

$$\begin{aligned}\frac{dx}{dt} &= \frac{k_2k_1a(a-x)}{k_{-1}a + k_2} \\ &= k'(a-x)\end{aligned} \tag{6.6}$$

So, on the basis of mechanism (6.1) and (6.2), the isomerisation reaction is predicted always to follow first order kinetics. References to "the second order region" of a unimolecular reaction are the unfortunate consequence of mis-applying the Lindemann mechanism by omitting step (6.2). There is simply no experimental basis for this terminology.

From equation (6.6), the first order rate constant, k', may be equated as follows:

$$k' = \frac{k_2k_1[A]_o}{k_{-1}[A]_o + k_2} \tag{6.7}$$

where $[A]_o$ denotes the initial concentration of A. At sufficiently high pressures, $k_{-1}[A]_o$ is much greater than k_2, so that to a good approximation the high pressure rate constant is given by:

$$k'_\infty = \frac{k_2k_1}{k_{-1}} = k_2K_1 \tag{6.7a}$$

* The comparability of k_1 and k_3 is a nice point. Usually k for a reaction between identical species is half that between otherwise distinguishable species (see p. 67). In this case, the k values are equal because in step 1, the excited molecule A* may be produced from either of the colliding molecules.

where K_1 is the equilibrium constant for step 1.

At appreciably lower pressures, $k_{-1}[A]_o$ is much less than k_2 so that we now have:

$$k'_{lp} = k_1[A]_o \tag{6.7b}$$

Thus the Lindemann theory predicts that at low pressures, the first order rate constant will decline from its high pressure limit and become proportional to the gas pressure employed in the experiment.

Equations (6.7a) and (6.7b) represent the two extreme situations arising from the mechanism (6.1) plus (6.2). Where the reversible reaction steps, 1/ −1 and 3/ −3, are in equilibrium, so that there is the equilibrium concentration of A^*, equation (6.7a) applies and the rate constant for step 2 is relevant. Equation (6.7b) represents a situation where the concentration of A^* is much less than the equilibrium value. The rate of reaction is now controlled by steps 1 and 3 and the rate constant for step 2 is immaterial. In a sense, the reaction no longer fits the label of a "unimolecular" reaction since the rate is governed by steps 1 and 3, both of which are bimolecular processes.

In order to represent the effect of added inert gases, it is necessary to add to the mechanism the reversible step,

$$A + M \underset{-4}{\overset{4}{\rightleftarrows}} A^* + M \tag{6.8}$$

where M denotes a molecule of an unreactive gas. This has the effect that equation (6.6) is replaced by,

$$\frac{dx}{dt} = k_2[A^*] = \frac{k_2\{k_1[A]_o + k_4[M]\}(a - x)}{k_{-1}[A]_o + k_{-4}[M] + k_2} \tag{6.9}$$

which indicates that the reaction will follow first order kinetics. The first order rate constant, k'', may be identified as

$$k'' = \frac{k_2\{k_1[A]_o + k_4[M]\}}{k_{-1}[A]_o + k_{-4}[M] + k_2} \tag{6.10}$$

The high pressure limit is attained when k_2 is insignificant in relation to $k_{-4}[M]$. If also the concentration of M is much greater than A is initially, then we have, for the limiting value:

$$k''_\infty = \frac{k_2 k_4}{k_{-4}} \tag{6.10a}$$

The ratio k_4/k_{-4} is equal to the ratio of the equilibrium concentration of A^* to that of A. It is thus equal to k_3/k_{-3} and so to K_1, so that the same high pressure limit should be attained in the presence of M as in its absence.

However, the success of the Lindemann theory is merely qualitative. From equation (6.7), one might expect that decreasing the pressure by two orders of magnitude should be sufficient to take one from the high pressure range where the rate constant is invariant to the low pressure

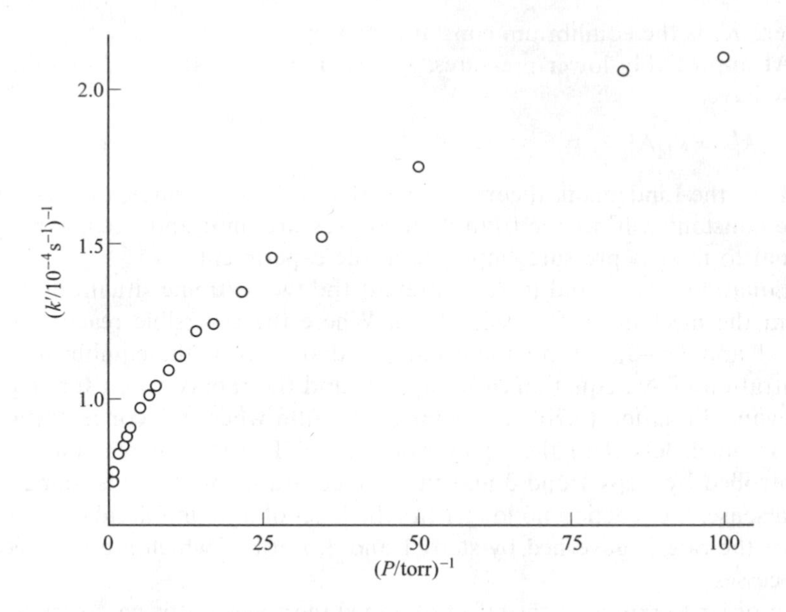

Fig. 6.2 Plot of the reciprocal of the first order rate constant against the reciprocal of the initial gas pressure for the isomerisation of 3-methylcyclobutene, at 396.0 K, using the data in Figure 6.1.

range where k' is proportional to the initial pressure. In practice, the transition is much more gradual and may span more than four orders of magnitude.

A detailed test of the Lindemann theory is readily made by inverting equation (6.7):

$$\frac{1}{k'} = \frac{k_{-1}}{k_2 k_1} + \frac{1}{k_1 [A]_o} \tag{6.11}$$

Thus, the theory predicts that the reciprocal of the first order rate constant should be a linear function of the initial concentration of the reactant. Such a plot is shown in Figure 6.2, whose appreciable curvature demonstrates that equation (6.7) is not strictly obeyed.

6.3 The RRK (or Kassel) theory

Although the Lindemann theory was found wanting in a quantitative sense, its predictions fit nearly all aspects of the kinetic behaviour of unimolecular gas reactions. The reasonable conclusion was, not that the theory was utterly wrong, but rather that it was not quite right and needed to be improved. The essence of its deficiency is easily stated. The theory assumes a unique energised reactant molecule, reacting with a fixed rate constant. It would be much more realistic to think of a whole range of energised molecules, vibrationally excited to varying extents, with the rate constant for the conversion into products being the greater the higher the excess vibrational energy of the molecule.

During the next decade, the Lindemann theory was revised along the lines indicated above. Very similar treatments were published independently by Rice and Ramsperger and by Kassel. In view of their shared ideas this advance is usually referred to as the RRK theory; since the slightly later version is arguably the better, it would not be inappropriate to label this development the Kassel theory.

Let us start from the assumption that an energised molecule reacts (i.e. it isomerises or dissociates) when the total energy in one vibrational mode, called the critical mode, exceeds some specified amount, ε^*. Where the total energies in all vibrations, as measured from their zero points, is less than ε^*, reaction is precluded. Where it exceeds ε^*, reaction can occur only if and when sufficient energy has flowed into the critical mode.

Vibrational motion is essentially harmonic but the potential energy curve, especially of a stretching vibration, deviates appreciably from a parabola, particularly at high excitation energies. As a consequence of this anharmonicity, in molecules that are vibrationally excited, energy flow must be possible between vibrational modes, which will thus behave as loosely coupled oscillators.

If a molecule with n vibrational modes has total vibrational energy ε, then the probability p that the amount of energy in the critical mode is at least ε^* is given by:

$$p = \left(\frac{\varepsilon - \varepsilon^*}{\varepsilon}\right)^{n-1} \tag{6.12}$$

This means that for $\varepsilon < \varepsilon^*$, p is zero. Above ε^*, p rises towards unity, but this rise is the slower the greater the value of n, as is illustrated in Figure 6.3. The derivation of equation (6.12) is given in Appendix A.

Since reaction is assumed necessarily to follow once energy ε^* is present in the critical mode, the effective rate constant for the reaction of the energised molecule, step 2 of mechanism (6.1), is given by the product of the probability p and the rate at which energy flows between oscillators. The latter is usually called ν and is of the order of magnitude of a vibrational frequency. So for such a molecule we have:

$$k_2 = \nu\left(\frac{\varepsilon - \varepsilon^*}{\varepsilon}\right)^{n-1} \tag{6.13}$$

At the high pressure limit of the first order rate constant, inter-molecular collisions are sufficiently frequent that the random distribution of vibrational energy is maintained. For a molecule with n vibrational modes, the fraction having total vibrational energy between ε and $\varepsilon + d\varepsilon$ is given by:

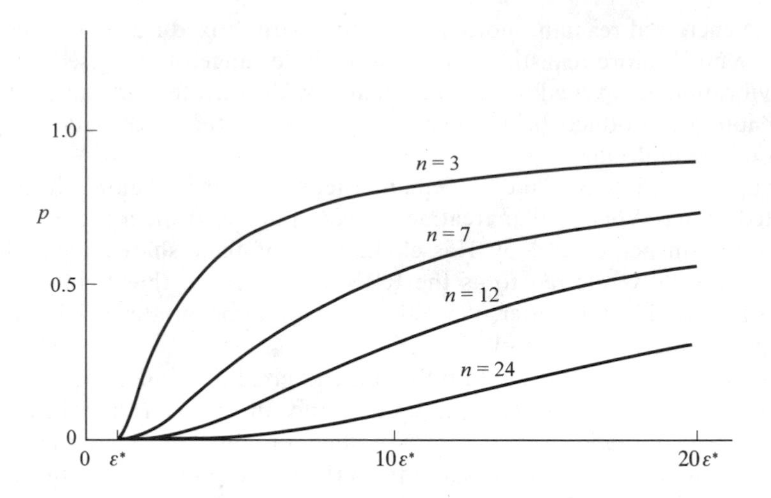

Fig. 6.3 Illustration of the dependence of p on ε for various values of n, based on equation (6.12).

$$dn_\varepsilon = \frac{1}{(n-1)!}\left(\frac{\varepsilon}{k_B T}\right)^{n-1} \frac{1}{k_B T}\, e^{-\varepsilon/k_B T}\, d\varepsilon \qquad (6.14)$$

Thus the high pressure rate constant, k'_∞, which we derived in equation (6.7a) as the product of the rate constant k_2 and the equilibrium constant K_1, relating to the two steps in mechanism (6.1), will be obtained from the product of the fraction of molecules capable of reacting and their rate constant for reaction. Both are functions of ε, so we need to integrate the product over the range of energies from ε^* to infinity.

$$\begin{aligned} k'_\infty &= k_2 K_1 \\ &= \int_{\varepsilon^*}^{\infty} \nu \left(\frac{\varepsilon - \varepsilon^*}{\varepsilon}\right)^{n-1} \frac{1}{(n-1)!}\left(\frac{\varepsilon}{k_B T}\right)^{n-1} \frac{1}{k_B T}\, e^{-\varepsilon/k_B T}\, d\varepsilon \qquad (6.15) \\ &= \nu e^{-\varepsilon^*/k_B T} \end{aligned}$$

(The above integral is also explained in Appendix A.)

In the fall-off region, the first order rate constant, given by equation (6.7) may be written:

$$k' = \frac{(k_2 k_1)/k_{-1}}{1 + (k_2/k_{-1}[A]_o)} \qquad (6.7c)$$

Substituting as before leads to the equation,

$$k' = \int_{\varepsilon^*}^{\infty} \frac{\nu\left(\dfrac{\varepsilon-\varepsilon^*}{\varepsilon}\right)^{n-1} \dfrac{1}{(n-1)!}\left(\dfrac{\varepsilon}{k_B T}\right)^{n-1} \dfrac{1}{k_B T}\, e^{-\varepsilon/k_B T}}{1+\dfrac{\upsilon}{k_{-1}[A]_o}\left(\dfrac{\varepsilon-\varepsilon^*}{\varepsilon}\right)^{n-1}}\, d\varepsilon$$

$$= \frac{\nu e^{-\varepsilon^*/k_B T}}{(n-1)!}\int_o^{\infty} \frac{y^{n-1} e^{-y}}{1+(\nu/k_{-1}[A]_o)\{y/(y+b)^{n-1}\}}\, dy \qquad (6.16)$$

where $y = (\varepsilon - \varepsilon^*)/k_B T$ and $b = \varepsilon^*/k_B T$.

This expression is capable of being evaluated by numerical methods. In this way, the dependence of k' on the initial gas concentration $[A]_o$ may be found, and so the theory may be compared with the experimental results.

At very low pressures, where the factor $k_{-1}[A]_o$ has become very small, equation (6.16) may be simplified.

$$k' = \frac{\nu e^{-\varepsilon^*/k_B T}}{(n-1)!} \frac{k_{-1}[A]_o}{\nu}\int_o^{\infty} (y+b)^{n-1} e^{-y}\, dy$$

$$= k_{-1}[A]_o\, e^{-\varepsilon^*/k_B T}\left[\left(\frac{\varepsilon^*}{k_B T}\right)^{n-1} \frac{1}{(n-1)!} + \ldots + 1\right] \qquad (6.16a)$$

One interesting feature of this equation is that k' is no longer showing the same temperature dependence as at the high pressure limit. Assuming that k_{-1} is proportional to $T^{1/2}$, we may evaluate the Arrhenius activation energy as follows:

$$E_a = -R\frac{d \ln k'}{d(1/T)}$$

$$= \frac{\varepsilon^* + \frac{1}{2}k_B T}{1+(n-1)\dfrac{k_B T}{\varepsilon^*} + \ldots + (n-1)!\left(\dfrac{k_B T}{\varepsilon^*}\right)^{n-1}} \qquad (6.17)$$

Since $\varepsilon^* \gg k_B T$, the terms in the denominator will get progressively smaller. Ignoring terms after the second, this gives, to an approximation:

$$E_a \approx \varepsilon^* - (n - \tfrac{3}{2})k_B T \qquad (6.17a)$$

Thus the RRK theory provides a quantitative explanation for the gradual decrease in the experimental activation energy of the reaction as a function of the initial pressure of the reactant.

The use of equation (6.16) to calculate k' at various initial concentrations of the reactant requires assumptions to be made in regard to the value of k_{-1}. It would seem reasonable that every collision can cause A^* to be de-energised, so that k_{-1} is in effect a collision number, as given in equation (4.2) of Chapter 4, and is proportional to $T^{1/2}$.

When an acceptable value has been chosen for the collision diameter d, and the necessary integrals evaluated, it is usually found that very poor agreement is obtained between experiment and calculation for the fall-off of k' with decreasing pressure of the reactant, which is illustrated in Figure 6.1. However, the degree of concurrence may be much improved if one does the calculation, not on the basis of the actual number of vibrational modes of the reactant molecule, but with a lesser number. The value of n which gives the best agreement with experiment is usually around half the actual number of vibrational modes of the reacting molecule.

This is readily exemplified using the data shown in Figure 6.1. Optimum agreement with these results was found when n was taken as 14: the actual number of vibrational modes of 3-methylcyclobutene is of course 33.

Despite this disparity, equation (6.16) was defended on the grounds that the lesser value of n probably represented the number of vibrational modes from which energy could easily flow to the critical mode. However, this plea is scarcely consistent with the earlier assumption, fundamental to RRK theory, that vibrational energy can flow freely between loosely coupled oscillators.

Moreover, since the RRK theory was proposed, experimental evidence has been obtained, in the course of several studies, to support the view that vibrational energy is readily transferred between modes. For example, vibrationally excited (or "hot") methylcyclopropane prepared by two different routes yielded on isomerisation the same mixture of butenes, even though the two sources gave hot molecules in which the excitation energy was initially in quite different vibrational modes.

6.4 The RRKM theory

The best theoretical treatment of unimolecular gas reactions currently available is based on that developed by Marcus, as an extension to and improvement of the RRK theory. Not surprisingly, this is often referred to as the RRKM theory. Its main achievements will now be outlined.

This theory considers the reaction sequence,

$$\left.\begin{array}{rcl} \mathrm{A+M} & \underset{-1}{\overset{1}{\rightleftarrows}} & \mathrm{A^*+M} \\ \mathrm{A^*} & \overset{2}{\rightarrow} & \mathrm{A^\dagger} \\ \mathrm{A^\dagger} & \overset{3}{\rightarrow} & \text{products} \end{array}\right\} \qquad (6.18)$$

where A^* denotes an energised A molecule, potentially capable of undergoing reaction, M represents any molecule present and capable of undergoing inter-molecular collisions with A and A^*, and $A^\dagger$ represents the transition state for the reaction process, as defined in Section 4.2.

At high pressures, where the interval between successive collisions is much less than the lifetime of A^*, there should be the steady state concentration of A^*, with an equilibrium distribution of species having various amounts of excess vibrational energy. An equilibrium between A^* and $A^\dagger$ is also assumed. The equilibrium constants may be expressed by statistical thermodynamics in terms of the relevant molecular partition functions. In order to work out the concentration of $A^\dagger$ using these equilibrium constants, it is necessary to make assumptions regarding the inter-atomic distances and the vibration frequencies of the transition state, so that the relevant rotational and vibrational partition functions may be evaluated.

In this way, an expression may be derived for the first order rate constant at the high pressure limit, namely:

$$k'_\infty = \ell^\dagger \frac{k_B T}{h} \frac{q_\dagger}{q_A} e^{-\varepsilon^*/k_B T} \tag{6.19}$$

Here $\ell^\dagger$ represents the statistical factor equal to the number of distinct reaction paths which are kinetically equivalent: for the dissociation of ethane into two methyl radicals this will be one, whereas for the dissociation of octafluoro-cyclobutane into two molecules of tetrafluoroethene, where the reaction may occur by either of two distinct but equivalent paths, it will be two. The factor $q_\dagger/q_A$ is the ratio of the partition functions of the transition state and the reactant molecule A, and ε^* is the minimum amount of vibrational energy which A^* must possess so that reaction is possible.

In equation (6.19), the factor k_BT/h has the value, at 298 K, of $10^{12.8}$, which is at the lower end of the range of vibrational frequencies, corresponding, in wave numbers, to a vibration at 207 cm^{-1} . The statistical factor $\ell^\dagger$ is never less than unity, but only rarely does it exceed 10. However, the presence of the factor $q_\dagger/q_A$ means that the pre-exponential factor for the high pressure rate constant is not constrained to a value close to 10^{13} s^{-1}. That was one of the uncomfortable implications of the RRK theory, as expressed by equation (6.15), where the factor ν was assumed to be approximately the same as the frequencies of the vibrational modes themselves. Experimentally, the pre-exponential factors of unimolecular reactions show appreciable variation from 10^{13} s^{-1} as is demonstrated in Table 6.1.

Table 6.1 High pressure Arrhenius parameters of some unimolecular gas reactions.

Reactant	Product(s)	$\log_{10}(A/\text{s}^{-1})$	$E_a/\text{kJ mol}^{-1}$
cis-2-propene	*trans*-2-propene	13.8	263
cyclobutane	1,3-butadiene	13.3	13.7
cyclopropane	propene	15.2	272
bromocyclobutane	1,3-butadiene + HBr	13.6	217
isopropenylcyclobutane	ethene + 1,3-butadiene	15.2	219
cyclopentane	1-pentene	16.1	355
cyclohexene	1-hexene	16.7	369
azomethane	$2CH_3^{\cdot}$ + N_2	17.2	232

Data taken from Lin, M. C. and K. J. Laidler, 1968, *Trans Faraday Soc.*, **64**, 94–102 and Holbrook, K. A., 1983, *Chem. Soc. Rev.*, **12**, 163–211.

In equation (6.19), the only significant aspect of the ratio, $q_{\ddagger}/q_{A}$, is that of the vibrational partition functions. Since A and the transition state have the same molecular weight and virtually the same shape, the translational partition functions are identical and the rotational partition functions cancel each other. The transition state has of course one less vibrational mode than has A: the "missing" mode corresponds to motion along the reaction co-ordinate. In evaluating equation (6.19) it is prudent to use, for the transition state, vibrational frequencies which seem reasonable and which will also lead to agreement between the calculated and experimental pre-exponential factors.

In other ranges of pressure, the RRKM theory leads to a more involved expression whose evaluation requires numerical integration techniques. It is not normally found that the choice of vibrational frequencies has much influence on the fit of the calculated to the experimental rate constant. Also, the form of these equations is such that it is not easily possible to obtain, even at quite low pressures, a simple equation for the activation energy. However, this may be determined by first evaluating the rate constant at a number of different temperatures, and these calculations show that the activation energy decreases as the rate constant falls off. For example, for the isomerisation,

$$\text{cyclo-}(CH_2\text{—}CH_2\text{—}CCl_2) \longrightarrow ClCH_2\text{—}CCl{=}CH_2 \tag{6.20}$$

for which E^* is taken as 232.2 kJ mol^{-1}, at 1000 torr, the Arrhenius activation energy is calculated as 241.4 kJ mol^{-1} but at 10^{-5} torr this has fallen to 203.3 kJ mol^{-1}. Thus the RRKM theory also is entirely consistent with this aspect of the experimental results.

A situation analogous to that of the vibrationally excited molecule, A^*, considered above arises in mass spectrometry where ionisation, for example by electron impact, generates a positive ion in a highly vibrationally excited state. Just as ethane may, at high temperatures, dissociate by the following routes,

$$\begin{aligned} C_2H_6^* &\rightarrow CH_3^{\cdot} + CH_3^{\cdot} \\ &\rightarrow C_2H_5^{\cdot} + H^{\cdot} \end{aligned} \tag{6.21}$$

so also the excited ethane ion may fragment:

$$\begin{aligned} (C_2H_6^{+})^* &\rightarrow CH_2CH_2^{+} + H_2 \\ &\rightarrow CH_3CH^{+} + H_2 \\ &\rightarrow C_2H_5^{+} + H^{\cdot} \\ &\rightarrow CH_3^{+} + CH_3^{\cdot} \end{aligned} \tag{6.22}$$

There are similarities between the positions of bond rupture occurring in these two instances, but the products of the excited ion may show more resemblance to those from vacuum–UV photolysis of ethane, where the molecules have been more highly excited than will readily occur by thermal excitation. The treatment of the ion fragmentation process, initially put forward in 1952 by Eyring and co-workers, called the Quasi-Equilibrium Theory, has many points of resemblance to the RRKM theory.

Suggested reading

Alvarino, J. M. and J. E. Figueruelo, 1977. The role of reaction products in unimolecular gas reactions. *Journal of Chemical Education,* **54**, 674–5.

Perlmutter-Hayman, B., 1967. Unimolecular gas reactions at low pressures. *Journal of Chemical Education*, **44**, 605–6.

Gilbert, R. G. and S. C. Smith, 1990. *Theory of Unimolecular and Recombination Reactions*. Oxford: Blackwell.

Robinson, P. J. and K. A. Holbrook, 1972. *Unimolecular Reactions*. London: Wiley-Interscience.

Laidler, K. J., 1969. *Theories of Chemical Reaction Rates*. New York: McGraw-Hill.

Marcus, R. A., 1952. Unimolecular dissociations and free radical recombination reactions. *Journal of Chemical Physics*, **20**, 359–64.

Wieder, G. M. and R. A. Marcus, 1962. Dissociation and isomerization of vibrationally excited species. II Unimolecular reaction rate theory and its application. *Journal of Chemical Physics*, **37**, 1835–52.

Lindemann, F. A., 1922. Discussion on the radiation theory of chemical reaction. *Transactions of the Faraday Society*, **17**, 598–9.

Holbrook, K. A., 1983. Current aspects of unimolecular reactions. *Chemical Society Reviews*, **12**, 163–211.

Tardy, D. C. and B. S. Rabinovitch, 1977. Intermolecular vibrational energy transfer in thermal unimolecular systems. *Chemical Reviews*, **77**, 369–408.

Problems

6.1 For the reaction,

$$\text{(1-methylcyclobutene, } CH_3 \text{ on ring)} \;=\; CH_2 : CH - \overset{\displaystyle CH_3}{\overset{|}{C}} : CH_2$$

in one experiment at 452.5 K the amount of 1-methylcyclobutene was found to decrease as follows:

t/min	3	5	7	10	15	20	25
% remaining	86.9	79.0	72.2	62.3	49.4	38.6	30.6

Determine the reaction order and evaluate the rate constant of this reaction. (Data from Frey, H. M., *Trans Faraday Soc.*, 1962, **58**, 957–60.)

6.2 For the isomerisation of 3-methylcyclobutene, at 421.6 K, the first order rate constant k' was found to vary with gas pressure p as follows:

p/torr	0.012	0.017	0.023	0.029	0.045	0.068
$k'/10^{-4}\,s^{-1}$	0.429	0.487	0.559	0.594	0.693	0.790

p/torr	0.091	0.148	0.214	0.437	0.613	1.10	6.49
$k'/10^{-4}\,s^{-1}$	0.858	0.969	1.04	1.17	1.21	1.30	1.46

Test whether these data are consistent with the Lindemann theory of unimolecular reactions. (Data from Frey, H. M. and D. C. Marshall, *Trans Faraday Soc.*, 1965, **61**, 1715–21.)

6.3 Spiropentane,

(where the planes of the three-membered rings are mutually perpendicular) is not stable and promptly rearranges to methylenecyclobutane

CH_2

A dideuterospiropentane is prepared in a vibrationally excited state by the reaction of dideuteromethylene with methylene cyclopropane:

$=CH_2$ + CD_2 = CD_2

If complete vibrational energy randomisation occurs before rearrangement takes place, what is the expected distribution of D atoms in the product methylenecyclobutane? (See Doering, W. von E., J. C. Gilbert and P. A. Leermakers, *Tetrahedron*, 1968, **24**, 6863–72.)

7 Chain reactions

In several types of chemical reactions, there is involved a species which is neither a reactant nor a product and which is usually called a reaction intermediate. Such a species (or perhaps, more than one such) is involved in a particular class of non-elementary reactions that have been given the name, "chain reactions". The essential characteristic of this group of reactions is that the succession of reaction steps which leads to the reactions product(s) also regenerates the reaction intermediate(s). In consequence, the kinetics of this group of reactions exhibit certain distinctive features.

7.1 Linear chain reactions

Having studied the kinetics of the reaction of H_2 with I_2 in the gas phase, referred to in Section 3.6, Bodenstein then carried out, with Lind, a similar study on the reaction of H_2 with Br_2. It was expected that the kinetics of this reaction would be similar, but the results did not confirm this. Rather, it was found that the order with respect to Br_2 was not one but one-half and that the forward reaction was inhibited by HBr. Their empirical equation for the rate, reported in 1906, is,

$$\frac{d[HBr]}{dt} = \frac{l[H_2][Br_2]^{\frac{1}{2}}}{1 + m[HBr]/[Br_2]} \tag{7.1}$$

where l and m are constants. Obviously this expression is not consistent with the elementary bimolecular process which might have been imagined to take place.

The explanation for this unusual expression for the reaction rate was not immediately apparent, but a satisfactory mechanism was proposed, independently, by three different scientists in 1919. It involved the newly developed concept of a chain reaction, introduced by Bodenstein himself in 1913 when he was trying to explain how hydrogen reacts with chlorine. The introduction of the name, chain reaction, has been credited to Christiansen of Copenhagen: in Danish it is "*kædereaktion*".

The mechanism proposed and widely accepted for the reaction of H_2 with Br_2 is as follows:

$$\left.\begin{array}{rcl} Br_2 & \xrightarrow{1} & Br^{\cdot} + Br^{\cdot} \\ Br^{\cdot} + H_2 & \xrightarrow{2} & HBr + H^{\cdot} \\ H^{\cdot} + Br_2 & \xrightarrow{3} & HBr + Br^{\cdot} \\ H^{\cdot} + HBr & \xrightarrow{4} & H_2 + Br^{\cdot} \\ Br^{\cdot} + Br^{\cdot} & \xrightarrow{5} & Br_2 \end{array}\right\} \qquad (7.2)$$

This mechanism illustrates the usual features of chain reactions. The species $Br^{\cdot}$ and $H^{\cdot}$ are the reaction intermediates which serve as chain carriers. Bromine atoms are formed in step 1 which is consequently called the initiation step. In each of steps 2 and 3, the participation of a chain carrier produces a molecule of product and the other chain carrier: thus these are called propagation steps. In step 5, two chain carriers are removed so this represents a termination step. These three types of reaction step, initiation, propagation and termination, are recurring features of a chain reaction.

Step 4 of mechanism (7.2) is more unusual. It is normally called a chain inhibition step and it achieves two things. It represents the reverse of step 2 and thus the removal of a molecule of product. Also, its occurrence means that the chain carrier $H^{\cdot}$ has been replaced by the chain carrier $Br^{\cdot}$ which may be removed in the termination step.

At first sight it may seem unusual that only one termination step is mentioned in the reaction scheme, since the steps,

$$\begin{array}{rcl} Br^{\cdot} + H^{\cdot} & \xrightarrow{6} & HBr \\ H^{\cdot} + H^{\cdot} & \xrightarrow{7} & H_2 \end{array}$$

would seem to be just as feasible as step 5 of (7.2). The dissociation energies of H_2, Br_2 and HBr are respectively 436, 193 and 366 kJ mol^{-1}, which means that, whereas step 3 is exothermic by 173 kJ mol^{-1}, step 2 is endothermic by 70 kJ mol^{-1} and must have an activation energy of at least that amount. In consequence, reaction 3, with a lower activation energy, is much faster than reaction 2, so that the concentration of H atoms will be very much smaller than that of Br atoms. Thus the only significant termination reaction is step 5 and steps 6 and 7 may be neglected.

All three of these termination reactions involve atom + atom recombination and will thus require a third body to help dispose of the excess vibrational energy (see Section 3.4). In that sense, all are presumably third order processes. For present purposes, this aspect may be set aside since the pressure of gas will remain constant throughout a reaction. However, strictly speaking, our second order rate constant for step 5, k_5, is really $k'_5[M]$, where k'_5 is the third order rate constant for the process:

$$Br^{\cdot} + Br^{\cdot} + M \rightarrow Br_2 + M \qquad (7.3)$$

In order to deduce the rate expression corresponding to a proposed chain reaction mechanism, it is necessary to invoke the Principle of Stationary States in regard to the chain carriers. So in the present case, we have:

$$\frac{d[Br]}{dt} = 2k_1[Br_2] - k_2[Br][H_2] + k_3[H][Br_2] + k_4[H][HBr] - 2k_5[Br]^2 = 0 \tag{7.4}$$

$$\frac{d[H]}{dt} = k_2[Br][H_2] - k_3[H][Br_2] - k_4[H][HBr] = 0 \tag{7.5}$$

The coefficients "2" in the first and last terms of equation (7.4) arise for slightly different reasons. There is a factor "2" in front of k_1 because reaction step 1 generates two Br atoms: that in regard to step 5 arises from the definition of the rate constant for a bimolecular reaction between identical species (see pages 3 and 67).

In this instance, the algebra is very simple. By adding equations (7.4) and (7.5) one obtains:

$$2k_1[Br_2] - 2k_5[Br]^2 = 0 \tag{7.6}$$

Thus

$$[Br^{\cdot}] = \left(\frac{k_1}{k_5}\right)^{1/2} [Br_2]^{1/2} \tag{7.7}$$

From equation (7.5), the concentration of $H^{\cdot}$ can be expressed in terms of that of Br:

$$[H^{\cdot}]\{k_3[Br_2] + k_4[HBr]\} = k_2[H_2][Br^{\cdot}] \tag{7.8}$$

$$\therefore [H] = \frac{k_2[H_2][Br^{\cdot}]}{k_3[|Br_2] + k_4[HBr]}$$

$$= \frac{k_2(k_1/k_5)^{1/2}[H_2][Br_2]^{1/2}}{k_3[Br_2] + k_4[HBr]} \tag{7.9}$$

In equations (7.7) and (7.9), the concentrations of the two chain carriers have now been obtained in terms of the rate constants of steps 1 to 5 and the concentrations of the reactants and product. These can now be substituted into the equation for the net rate of formation of the reaction product, to yield the desired expression:

$$\frac{d[HBr]}{dt} = k_2[Br][H_2] + k_3[H][Br_2] - k_4[H][HBr]$$

$$= k_2\left(\frac{k_1}{k_5}\right)^{1/2}[H_2][Br_2]^{1/2} +$$

$$k_3[Br_2] - k_4[HBr]\} \frac{k_2(k_1/k_5)^{1/2}[H_2][Br_2]^{1/2}}{k_3[Br_2] + k_4[HBr]}$$

$$= k_2\left(\frac{k_1}{k_5}\right)^{1/2}[H_2][Br_2]^{1/2}\left\{1 + \frac{k_3[Br_2] - k_4[H_2]}{k_3[Br_2] + k_4[H_2]}\right\}$$

$$= \frac{2k_2\left(\frac{k_1}{k_5}\right)^{1/2}[H_2][Br_2]^{1/2}}{1 + \frac{k_4[HBr]}{k_3[Br_2]}} \qquad (7.10)$$

This equation implies the identical dependence of the reaction rate on the concentrations to that of Bodenstein and Lind, equation (7.1). Thus l is equal to $2k_2(k_1/k_5)^{1/2}$ and m to k_4/k_3.

The reaction of H_2 with I_2 was referred to in Chapter 3 and exhibits second order kinetics. The major difference between the reactions of bromine and iodine with hydrogen is in regard to the step involving X and H_2: does it yield HX and $H^{\cdot}$? In the case of the I atom, the H–I bond is so weak (299 kJ mol^{-1}) that this outcome is so endothermal (by 137 kJ mol^{-1}) that the activation energy (which must exceed this figure) is prohibitive. The Br atom can achieve this outcome, but not easily, so Br_2 and H_2 combine by a chain reaction. For the chlorine atom there is no difficulty, but of course the Cl_2 molecule is less easily dissociated.

7.2 Pyrolysis reactions involving chain mechanisms

A number of pyrolyses of organic compounds have been shown to involve chain reactions, following a mechanism of the type proposed by Rice and Herzfeld. These normally exhibit a kinetic order of three-halves. One example is provided by the thermal decomposition of ethanal (acetaldehyde), where the major products are methane and carbon monoxide,

$$CH_3CHO \;=\; CH_4 + CO \qquad (7.11)$$

with minor amounts of ethane and hydrogen, and of some other products.

The mechanism proposed for this reaction is as follows:

$$\left.\begin{array}{rcl} CH_3CHO & \xrightarrow{1} & CH_3^{\cdot} + CHO^{\cdot} \\ CHO^{\cdot} & \xrightarrow{2} & CO + H^{\cdot} \\ H^{\cdot} + CH_3CHO & \xrightarrow{3} & H_2 + CH_3CO^{\cdot} \\ CH_3^{\cdot} + CH_3CHO & \xrightarrow{4} & CH_4 + CH_3CO^{\cdot} \\ CH_3CO^{\cdot} & \xrightarrow{5} & CH_3^{\cdot} + CO \\ CH_3^{\cdot} + CH_3^{\cdot} & \xrightarrow{6} & C_2H_6 \end{array}\right\} \tag{7.12}$$

Here, step 1 represents initiation, producing two radical species. The formyl radical ($^{\cdot}CHO$) is quantitatively converted, in two steps, into the acetyl radical ($CH_3CO^{\cdot}$). The propagation reactions are thus steps 4 and 5, involving the chain carriers $CH_3^{\cdot}$ and $CH_3CO^{\cdot}$: each step produces a molecule of one of the products and the other chain carrier. Step 6 represents mutual termination and is the only important termination step because the concentration of methyl radicals is much greater than that of acetyl, reflecting the fact that step 4 has an appreciable activation energy.

To work out the rate expression, it is convenient to start with the radical species $CHO^{\cdot}$ and $H^{\cdot}$, which are reactive intermediates but not chain carriers. Applying the Principle of Stationary States we have:

$$\frac{d[CHO^{\cdot}]}{dt} = k_1[CH_3CHO] - k_2[CHO^{\cdot}] = 0 \tag{7.13}$$

$$\text{Thus, } [CHO^{\cdot}] = \frac{k_1}{k_2}[CH_3CHO] \tag{7.14}$$

$$\frac{d[H]}{dt} = k_2[CHO^{\cdot}] - k_3[H][CH_3CHO] = 0 \tag{7.15}$$

From equations (7.14) and (7.15) we have:

$$[H] = k_1/k_3 \tag{7.16}$$

The equations obtained from the chain carriers are:

$$\frac{d[CH_3CO]}{dt} = k_4[CH_3][CH_3CHO] + k_3[H][CH_3CHO] - k_5[CH_3CO] = 0 \tag{7.17}$$

$$\frac{d[CH_3]}{dt} = k_1[CH_3CHO] - k_4[CH_3][CH_3CHO] + k_5[CH_3CO] - 2k_6[CH_3]^2 = 0 \tag{7.18}$$

The concentration of H atoms in equation (7.17) may be removed by substituting from equation (7.16), giving:

$$\frac{d[CH_3CO]}{dt} = k_4[CH_3][CH_3CHO] + k_1[CH_3CHO] - k_5[CH_3CO] = 0 \tag{7.17a}$$

This result could have been obtained by inspection: every $CHO^{\cdot}$ formed in step 1 yields an H atom which yields an acetyl radical, so the rate of step 3 must be equal to that of step 1.

By adding equations (7.18) and (7.17a) we obtain,

$$2k_1[CH_3CHO] - 2k_6[CH_3^{\cdot}]^2 = 0 \tag{7.19}$$

which leads to:

$$[CH_3^{\cdot}] = \left(\frac{k_1}{k_6}\right)^{1/2}[CH_3CHO]^{1/2} \tag{7.20}$$

The rate of formation of methane is readily obtained:

$$\begin{aligned}\frac{d[CH_4]}{dt} &= k_4[CH_3^{\cdot}][CH_3CHO] \\ &= k_4\left(\frac{k_1}{k_6}\right)^{1/2}[CH_3CHO]^{3/2}\end{aligned} \tag{7.21}$$

Thus, the mechanism is shown to be consistent with a reaction order of 3/2. The three-halves order rate constant is equal to $k_4(k_1/k_6)^{1/2}$, so the Arrhenius activation energy may be equated to $\{E_4 + ½(E_1 - E_6)\}$. E_4 has been determined as 35.1 kJ mol^{-1}, E_1 as 336 kJ mol^{-1} and E_6 should be zero, so a value of 203.1 kJ mol^{-1} is expected. The experimental value of 205.4 kJ mol^{-1} is in very good agreement.

However, in some studies of acetaldehyde pyrolysis, the reaction order seemed to be close to two. This was especially the case if the order were determined from the manner in which the rate slowed down as the reaction was followed virtually to completion, and means that the pyrolysis decelerates to a greater extent than is expected for a three-halves order reaction.

Two other facts also show that reaction scheme (7.12), while substantially correct, is not the full story. Firstly, among the products of the pyrolysis are small amounts of acetone, propionaldehyde, ethene and carbon dioxide, none of which is accounted for by this scheme. Secondly, there is the evidence of the experiment in which a mixture of equimolar amounts of CH_3CHO and CD_3CDO was pyrolysed. Scheme (7.12) requires that CH_4, CH_3D, and CHD_3 and CD_4 should be formed, and they were all found to be present. Not surprisingly, in view of the expected primary kinetic isotope effect in step 4, there was more CH_4 than CH_3D and more CHD_3 than CD_4. However, an appreciable amount of CH_2D_2 was also observed, for whose presence no simple explanation can be supplied on the basis of mechanism (7.12).

Additional reaction steps that have been proposed include the following:

$$CH_3^{\cdot} + CH_3CHO \rightarrow CH_4 + \dot{C}H_2CHO \tag{7.22}$$

$$CH_3^{\cdot} + CH_2CHO \rightarrow CH_3CH_2CHO \tag{7.23}$$

$$CH_3^{\cdot} + CH_3CHO \rightarrow CH_3CO\,CH_3 + H^{\cdot} \tag{7.24}$$

which can account for the small yields of acetone and propionaldehyde.

As the pyrolysis reaction proceeds, the accumulation of products can easily influence the chain termination. An additional factor affecting the kinetics is that the rate constant of step 1 of scheme (7.12), which is unimolecular dissociation, is expected to fall off at low pressures in the manner discussed in Chapter 6.

7.3 Chain polymerisation: free radical processes

Chain (or addition) polymerisation may be represented by the overall equation,

$$nCH_2 : CHX \;=\; -\!\!(CH_2CHX)\!\!-_n \tag{7.25}$$

There are three different types of mechanism, depending on whether the (kinetic) chain carriers are free radicals, anions or cations. Free radical chain polymerisation mechanisms show some resemblance to those discussed in Sections 7.1 and 7.2. They require, as initiator, a compound capable of producing free radicals by either a thermal or a photochemical reaction. Examples of the former are benzoyl peroxide and azo-*bis*-isobutyronitrile:

$$\left.\begin{array}{lcl}(PhCOO)_2 & \rightarrow & PhCOO^{\cdot} + PhCOO^{\cdot} \\ PhCOO\cdot & \rightarrow & Ph^{\cdot} + CO_2\end{array}\right\} \tag{7.26}$$

and

$$\underset{\displaystyle CN}{\underset{|}{Me_2C}}{-}N{=}N{-}\underset{\displaystyle CN}{\underset{|}{C\,Me_2}} \longrightarrow 2\,\underset{\displaystyle CN}{\underset{|}{Me_2C\cdot}} + N_2 \tag{7.27}$$

In general, we may write the initiation process as,

$$I \xrightarrow{k_i} R\cdot + R\cdot \tag{7.28}$$

where I represents the initiator species and R· the free radical resulting from its dissociation in the initiation step. Some fraction, *f*, of the free radical species so formed will react with a molecule of monomer, CH_2:CHX.

$$R\cdot + CH_2{:}CHX \rightarrow RCH_2\dot{C}HX \tag{7.29}$$

The first propagation step is then:

$$RCH_2\dot{C}HX + CH_2{:}CHX \rightarrow RCH_2CHX\,CH_2\dot{C}HX \tag{7.30}$$

This may more conveniently be written,

$$m^{\cdot}_1 + M \rightarrow m^{\cdot}_2 \tag{7.30a}$$

where M represents a molecule of the monomer and $m^{\cdot}_x$ the growing radical with precisely x monomer units incorporated. The general propagation step is,

$$m^{\cdot}_x + M \xrightarrow{k_p} m^{\cdot}_{x+1} \tag{7.31}$$

and the simplifying supposition is made that the rate constant for the propagation is independent of the value of x.

In the absence of impurities which might react with these free radicals, mutual termination will prevail. Two growing polymer chains may react either by combination or by disproportionation:

$$\left.\begin{aligned} m^{\cdot}_x + m^{\cdot}_y &\xrightarrow{k_{tc}} P_{x+y} \\ m^{\cdot}_x + m^{\cdot}_y &\xrightarrow{k_{td}} P_x + P_y \end{aligned}\right\} \tag{7.32}$$

These alternative reactions differ in regard to the molecular weight of the resulting polymer, but they are kinetically equivalent in that both represent the loss of two free radical centres. We will use k_t as the aggregate rate constant of the termination step, and it also is assumed independent of x.

Taking the rate of initiation as that of forming $RCH_2\dot{C}HX$, we have,

$$\text{Rate of initiation} = 2fk_i[I] \tag{7.33}$$

where I represents the initiator and k_i the first order rate constant for its dissociation. The factor f will lie between zero and one.

This may be equated to the rate of disappearance of free radical centres by termination step (7.32). Thus we have:

$$2fk_i[I] = 2k_t[m^{\cdot}_x]^2 \tag{7.34}$$

So, for the steady state concentration of the free radical centres we obtain:

$$[m^{\cdot}_x] = \left(\frac{fk_i}{k_t}\right)^{1/2}[I]^{1/2} \tag{7.35}$$

Substituting this concentration into the equation for the rate of the propagation step we obtain an expression for the rate of polymerisation, R_p:

$$\begin{aligned} R_p = -\frac{d[CH_2{:}CHX]}{dt} &= k_p[m^{\cdot}_x][M] \\ &= k_p\left(\frac{fk_i}{k_t}\right)^{1/2}[I]^{1/2}[M] \end{aligned} \tag{7.36}$$

Thus a free radical addition polymerisation process will be first order with respect to monomer concentration and one-half order with respect to the initiator.

The Arrhenius activation energy of the polymerisation reaction is readily obtained as

$$E_a = E_p + \tfrac{1}{2}(E_i - E_t) \tag{7.37}$$

where E_p, E_i and E_t are respectively the activation energies of the propagation, initiation and termination steps. This last reaction is frequently diffusion-controlled, in which case E_t will be very small.

It is, of course, possible to study the decomposition of initiator I in the absence of monomer, and thus to determine k_i. Combining this with the apparent three-halves order rate constant for the polymerisation reaction will lead to $k_p/k_t^{1/2}$. Special techniques are required in order to evaluate either of these individually.

It is appropriate to mention the effect of adding, to the polymerisation system described above, a compound such as a thiol, R′SH, containing a hydrogen atom attached by a bond of much lower bond dissociation energy. This can react fairly readily with a growing free radical centre:

$$\begin{aligned} & R'SH + R(CH_2{-}CHX)_xCH_2\dot{C}HX \\ & \qquad \rightarrow R'S^{\cdot} + R(CH_2{-}CHX)_xCH_2CH_2X \end{aligned} \tag{7.38}$$

The consequence is that, by acquiring an H atom from the thiol, there is no longer a free radical centre on the polymer chain. However, the $R'S^{\cdot}$ radical so generated is capable of acting in the same manner as the free radicals formed from the initiator:

$$R'S^{\cdot} + CH_2{:}CHX \rightarrow R'SCH_2\dot{C}HX \tag{7.39}$$

leading to a new free radical centre. Thus R′SH acts as a chain transfer reagent. Its presence has very little effect on the rate of consumption of the monomer, but it causes a decrease in the average length of the polymer molecules.

7.4 Chain polymerisation: ionic processes

Polymerisation of a monomer was shown to occur in cases where no recognised free radical initiator was present. In some instances it was found that the growing polymer was an anion, in some a cation. However, these processes tend to be much more specific than free radical polymerisation, in which virtually any olefin may engage. Cationic initiation occurs much more readily if there is an electron-donating substituent on the olefin, for example, an alkoxy group as in a vinyl ether, $CH_2{:}CH{-}OR$, and anionic initiation where there are electron-withdrawing substituents, as there are in acrylonitrile, $CH_2{:}CHCN$, or methyl methacrylate, $CH_2{:}C(CH_3)CO_2CH_3$. Styrene, $CH_2{:}CH\,Ph$, may engage in either process.

The presence of an ion means the existence of a counter-ion. These are more readily separated in liquids of high permittivity, but many of these are ineligible as solvents because they react with and destroy the ionic initiator. So in many instances these reactions may be studied only in media of low permittivity where strong ion-pairing occurs. The exclusion of impurities tends to be even more important than it is for free radical polymerisation reactions.

7.4.1 Anionic polymerisation

A known anionic polymerisation reaction is that of styrene in liquid ammonia, initiated by potassamide, KNH_2:

$$KNH_2 \rightleftharpoons K^+ + NH_2^- \tag{7.40}$$

The following mechanism has been proposed:

$$\left.\begin{aligned} NH_2^- + CH_2{:}CH\,Ph &\rightarrow H_2NCH_2^-CH\,Ph \\ &\qquad (m_1^-) \\ m_1^- + CH_2{:}CH\,Ph &\rightarrow m_2^- \\ m_x^- + CH_2{:}CH\,Ph &\rightarrow m_{x+1}^- \end{aligned}\right. \tag{7.41}$$

(Here we are using m_x^- to denote the anion incorporating precisely x molecules of monomer.)

The most important difference between an ionic polymerisation process and a free radical one is that, in the former, mutual termination simply cannot and does not occur. Thus, growing polymer anions may freely coexist without any fear of mutual annihilation.

In this system, termination may be caused by an impurity such as O_2, which underlines the need for the total exclusion of air. In the absence of any foreign molecules, termination may be achieved by proton transfer from the solvent:

$$m_x^- + NH_3 \rightarrow H_2N\,(CH_2CH\,Ph)_xH + NH_2^- \tag{7.42}$$

For kinetic purposes, the mechanism can be summarised:

$$\left.\begin{aligned} NH_2^- + M &\xrightarrow{k_i} m_1^- \\ m_x^- + M &\xrightarrow{k_p} m_{x+1}^- \\ m_x^- + NH_3 &\xrightarrow{k_t} P_x + NH_2^- \end{aligned}\right\} \tag{7.43}$$

Equating the rate of initiation to that of termination we obtain:

$$k_i\,[NH_2^-][M] = k_t[m_x^-][NH_3] \tag{7.44}$$

This leads to the following expression for the steady state concentration of growing polymer anions:

$$[m_x^-] = \frac{k_i[NH_2^-][M]}{k_t[NH_3]} \tag{7.45}$$

and thus an equation for the rate of polymerisation:

$$R_p = \frac{-d[CH_2{:}CHPh]}{dt} = k_p[m_x^-][M]$$
$$= \frac{k_p k_i[NH_2^-][M]^2}{k_t[NH_3]} \tag{7.46}$$

So, on the basis of mechanisms (7.37) plus (7.38), this polymerisation reaction is second order in monomer. The reaction is first order in the active initiator, NH_2^-, but the rate is more nearly proportional to the square root of the nominal concentration of potassamide, since this compound has a very low dissociation constant in liquid ammonia. The observed rate constant for the reaction may be equated to $(k_p k_i/k_t)$, so that the activation energy of this polymerisation, reported as 38 kJ mol^{-1}, is equal to $(E_p + E_i - E_t)$.

Another initiator is butyl lithium, where Bu^- plays the role in initiation corresponding to that of NH_2^- in mechanism (7.43). The polar molecule, $^{\delta-}Bu\text{–}Li^{\delta+}$ undergoes aggregation in media of low permittivity such as tetrahydrofuran (THF), where it exists mostly as the hexamer. Consequently, as is mentioned in Section 3.6, the rate of polymerisation is proportional to the nominal initiator concentration to the one-sixth power.

The initiator system, sodium *plus* naphthalene, may serve to illustrate the behaviour of the growing polymer anions. On adding sodium to naphthalene in THF, where oxygen and water have been rigidly excluded from both, a green colour is obtained from the naphthalenide radical anion:

$$\text{Na} + \text{Naphth} \rightleftharpoons \text{Na}^+ \ \text{Naphth}^{\overline{\cdot}} \tag{7.47}$$

If styrene is now added under the same conditions, electron transfer occurs to yield styrene$^{\overline{\cdot}}$, which is deep red in colour. Styrene$^{\overline{\cdot}}$, being a radical anion, may undergo simultaneous propagation reactions at each end, at one end as an anion and at the other as a free radical. However, the free radical ends quickly dimerise while the anionic ends continue to add monomer units in the manner of the propagation step of mechanism (7.43). In this system, termination of the anionic centres occurs to a negligible extent and the red colour of the "living polymer" anions persists long after the supply of monomer has been exhausted.

7.4.2 Cationic polymerisation

The formation of the positively charged species necessary for cationic polymerisation may be achieved by an acid reacting with a monomer molecule:

$$CH_2{:}CHX + H_2SO_4 \rightleftharpoons CH_3{-}\overset{+}{C}HX + HSO_4^- \qquad (7.48)$$

While an electron-deficient molecule cannot perform this, it may achieve the same result if it is possible to abstract a hydride ion from the monomer molecule:

$$CH_2{:}C(CH_3)_2 + SnCl_4 \rightleftharpoons CH_2{:}C(\overset{+}{C}H_2)(CH_3) + Sn\bar{C}l_4H \qquad (7.49)$$

However, if the monomer has no such reactive hydrogen atoms, then the electron-deficient molecule needs to be converted into a source of protons: a simple molecule like HCl or H_2O may serve as a co-initiator:

$$BF_3 + H_2O \rightleftharpoons H^+BF_3OH^- \qquad (7.50)$$

$$H^+BF_3OH^- + CH_2{:}CHX \rightleftharpoons CH_3\overset{+}{C}HX + BF_3OH^- \qquad (7.51)$$

The above examples illustrate two of the complexities of the kinetics of cationic polymerisation. Firstly, the initiation step rarely goes to completion, but rather it approaches an equilibrium. Secondly, the initiation process generates an ion pair and in media of low permittivity, ion pairing persists through the propagation stages:

$$(CH_2{-}CHX)_x{-}CH_2\overset{+}{C}HX\ A^- + CH_2{:}CHX \rightarrow (CH_2{-}\ CHX)_{x+1}{-}CH_2\overset{+}{C}HX\ A^- \qquad (7.52)$$

Given that X is a suitably electron-donating substituent, the propagation steps are usually extremely rapid, even at temperatures as low as –80°C. As with anionic polymerisation, mutual termination is ruled out. Since the carbocations involved are much more reactive than the corresponding free radicals or even the carbanions, reaction with an impurity may be an important termination route except under conditions of extreme purity. Another is reaction with the accompanying counter-ion, analogous to the reverse of reaction (7.48) or variations on it:

$$\sim\!\sim (CH_2{-}CHX)_x{-}CH_2\overset{+}{C}HX\ A^- \rightarrow \sim\!\sim (CH_2{-}CHX)_x{-}CH{:}CHX + HA \qquad (7.53)$$

Chain transfer reactions, involving proton transfer to a monomer molecule, can also occur. They do not represent the loss of the chain centre but they tend to restrict the molecular weight of the polymer.

Treating the kinetics of cationic polymerisation involves the difficulty that an equilibrium concentration of active centres is not readily achieved. Thus the ubiquitous tool of chain reaction kinetics, the Principle of Stationary States, cannot readily be applied. One expects the concentration of cationic centres to be proportional to the concentrations of initiator and of monomer. This would lead to the rate of propagation being proportional to the concentration of the initiator and to the square of the concentration of monomer, just as in anionic polymerisation.

In chain reactions, a useful parameter is the mean kinetic chain length, $\bar{x}_n$, which is the mean number of propagation steps in which a chain carrier engages before termination occurs. This is especially important in polymerisation reactions in that, in the absence of chain transfer reactions, the number of monomer units with which a chain centre reacts becomes imprinted within the product.

For cationic polymerisation, with chain termination as outlined above, we have:

$$\bar{x}_n = \frac{\text{Rate of propagation}}{\text{Rate of termination}} = \frac{k_p[m_x^+][M]}{k_t[m_x^+]}$$
$$= \frac{k_p}{k_t}[M] \qquad (7.54)$$

where m_x^+ represents a cationic centre, M a molecule of monomer and k_p and k_t are the rate constants for propagation and termination. Thus the Arrhenius activation energy for the mean kinetic chain length is $(E_p - E_t)$, which is expected to be negative since propagation occurs so readily, with a very low activation energy. This implies that lowering the temperature will lead to longer polymer chains, which has been shown experimentally.

This contrasts with the result obtained for free radical polymerisation. From equations (7.35) to (7.37) we have:

$$\bar{x}_n = \frac{k_p[m_x^{\cdot}][M]}{2k_t[m_x^{\cdot}]^2}$$
$$= k_p\left(\frac{1}{k_t k_i}\right)^{1/2}\left(\frac{1}{4f}\right)^{1/2}[M][I]^{-1/2} \qquad (7.55)$$

This gives the Arrhenius activation energy as $\{E_p - \frac{1}{2}(E_t + E_i)\}$, which is rarely negative since E_t is very small, so that here the kinetic chain length tends to become longer as the temperature is increased.

7.5 Less simple linear chain reactions

In the discussion of linear chain reactions to date, manipulating the equations obtained by applying the Principle of Stationary States has been a very simple matter. However, it is not inevitable that these equations can be so readily handled.

An illustration of the difficulties that may arise is provided by the decomposition of hydrogen peroxide in acidic solution in the presence of small amounts of copper(II) ion. This reaction may be initiated photochemically and, as is discussed in Chapter 11, the kinetics of photochemical reactions have their innate complications. However, the basic difficulty in applying the Principle of Stationary States to this system does not in any sense arise from the photochemical nature of the initiation step.

By illuminating a solution of hydrogen peroxide with UV light, H_2O_2 is dissociated into hydroxyl radicals and decomposition occurs, according to the equation:

$$2H_2O_2 \; = \; 2H_2O + O_2 \tag{7.56}$$

The following mechanism has been proposed:

$$\left.\begin{array}{rcl} H_2O_2 & \xrightarrow{h\nu} & OH + OH \\ OH + H_2O_2 & \xrightarrow{1} & H_2O + HO_2 \\ HO_2 + H_2O_2 & \xrightarrow{2} & O_2 + H_2O + OH \\ HO_2 + HO_2 & \xrightarrow{3} & H_2O_2 + O_2 \\ HO_2 + OH & \xrightarrow{4} & H_2O + O_2 \end{array}\right\} \tag{7.57}$$

Here, steps 1 and 2 represent propagation and steps 3 and 4 are alternative termination processes.

If a small amount of copper(II) ion is added, the rate of the photochemical decomposition is much enhanced, because propagation step 2, which is a rather difficult and slow step, is replaced. There is also a new termination step:

$$\left.\begin{array}{rcl} HO_2 + Cu^{2+} & \xrightarrow{5} & O_2 + H^+ + Cu^+ \\ Cu^+ + H_2O_2 & \xrightarrow{6} & Cu^{2+} + OH^- + OH \\ Cu^+ + HO_2 & \xrightarrow{7} & Cu^{2+} + HO_2^-(\xrightarrow{H^+} H_2O_2) \end{array}\right\} \tag{7.58}$$

It is expected that, except at extremely low concentrations of copper(II) ion, step 7 will predominate over step 4.

The rate of direct photolysis of H_2O_2 in the initial step of scheme (7.57) may be expressed as the product of I_a, the rate of light absorption in the solution (in Einstein $dm^{-3}\,s^{-1}$) and ϕ, the quantum yield (molecule quantum^{-1}). The net rate of decomposition of hydrogen peroxide is then:

$$-\frac{d[H_2O_2]}{dt} = \phi I_a + k_1[OH][H_2O_2] + k_6[Cu^+][H_2O_2] - k_7[Cu^+][HO_2] \tag{7.59}$$

Applying the Principle of Stationary States to each of the three chain carriers leads to the following equations:

$$\left.\begin{aligned}\frac{d[OH]}{dt} &= 2\phi I_a - k_1[OH][H_2O_2] + k_6[Cu^+][H_2O_2] = 0\\ \frac{d[HO_2]}{dt} &= k_1[OH][H_2O_2] - k_5[HO_2][Cu^{2+}] - k_7[HO_2][Cu^+] = 0\\ \frac{d[Cu^+]}{dt} &= k_5[HO_2][Cu^{2+}] - k_6[Cu^+][H_2O_2] - k_7[HO_2][Cu^+] = 0\end{aligned}\right\} \tag{7.60}$$

Adding these three equations leads to,

$$2\phi I_a - 2k_7[Cu^+][HO_2] = 0 \tag{7.61}$$

from which we have:

$$[Cu^+][HO_2] = \phi I_a/k_7 \tag{7.62}$$

However, this does not serve to identify the concentration of either of these chain carriers, nor that of the third one, the OH radical.

If it may be assumed that the chains are very long, so that the rate of the termination step is an insignificant fraction of that of each propagation step, then as an approximation, one may equate the rates of the propagation steps involving each of the chain carriers, viz.:

$$k_1[OH][H_2O_2] \approx k_5[HO_2][Cu^{2+}] \approx k_6[Cu^+][H_2O_2] \tag{7.63}$$

This, in conjunction with equation (7.62), permits the steady state concentrations of all three species to be evaluated and thus, within the limitations of the above assumption, the rate expression can be obtained.

It is more satisfactory to obtain a general solution. This requires more extensive algebra than has been required in the previous examples and is presented in Appendix B. The answer is that when the concentrations of the reaction intermediates are removed from equation (7.59), one has:

$$-\frac{d[H_2O_2]}{dt} = \phi I_a\left[1 + \frac{4k_5k_6[H_2O_2][Cu^{2+}]}{\phi I_a k_7}\right]^{1/2} \tag{7.64}$$

Under conditions of low light intensity and a substantial concentration of Cu^{2+}, the numerator of the last term far exceeds the denominator, so that, to a good approximation,

$$-\frac{d[H_2O_2]}{dt} = 2\left(\frac{\phi I_a k_5 k_6}{k_7}\right)^{1/2}[H_2O_2]^{1/2}[Cu^{2+}]^{1/2} \tag{7.65}$$

which is the answer obtained directly by accepting the approximation in (7.63) above.

Where the combination of the wavelength of the exciting light, the path length of the photolysis cell and the concentration of H_2O_2 is such that the absorbance of the cell contents at the exciting wavelength is much less than one, the rate of light absorption I_a is proportional to the concentration of H_2O_2 (see p. 221). Consequently, the rate of decomposition under these conditions is proportional to the first power of the concentration of H_2O_2, and to the one-half power of the copper(II) ion concentration.

7.6 Branching chains and explosion phenomena

In the chain reactions considered to date, each propagation step has involved one chain carrier as a reactant and one chain carrier, either similar or dissimilar, as a product. Under these circumstances, changes to the number of chain carriers are brought about only by the initiation and propagation steps.

It is possible to envisage a regime with less rigid demarcation, where the formation of the reaction product occurs by a step in which the number of chain carriers is increased. The term normally used to denote this is chain *branching*.

To consider the concentration of chain carriers, let us assume that there is only one such, $R^{\cdot}$, formed from an initiator I by a unimolecular process of rate constant k_i. Also, assume $R^{\cdot}$ undergoes linear termination, with a rate constant k_t. Let α represent the number of chain carriers produced in each chain branching step, of rate constant k_b, where M represents the reactant that engages in the propagation/branching step.

Applying the Principle of Stationary States, we then have:

$$\frac{d[R^{\cdot}]}{dt} = k_i[I] + (\alpha - 1)k_b[R^{\cdot}][M] - k_t[R^{\cdot}] = 0 \qquad (7.66)$$

$$\therefore [R^{\cdot}] = \frac{k_i[I]}{k_t - (\alpha - 1)k_b[M]} \qquad (7.67)$$

Given that $\alpha > 1$, there is a value of the concentration of M for which the denominator will become zero. As this is approached, the concentration of $R^{\cdot}$ and thus the rate of the propagation reaction tend to infinity. This means that an explosion takes place.

Thus two conditions are necessary for a chain reaction to cause an explosion. Chain branching must occur, but in addition, it is necessary that chain termination is linear, by a step involving only one chain carrier, rather than mutual, by a step involving two. Two examples of branching chain reactions will now be considered.

7.6.1 The hydrogen–oxygen reaction

Every schoolchild knows that hydrogen and oxygen combine to form water:

$$2H_2 + O_2 = 2H_2O \qquad (7.68)$$

When water is formed from its elements, this does not take place in a single termolecular reaction. Rather it occurs by a chain process, involving the radical species H, OH and O. The hydroxyl radical acts as a well-behaved chain carrier, participating in the step

$$OH + H_2 \rightarrow H_2O + H \tag{7.69}$$

which yields product plus another chain carrier. However, the major reactions of the other two species are chain-branching processes:

$$H + O_2 \rightarrow OH + O \tag{7.70}$$

and $$O + H_2 \rightarrow OH + H \tag{7.71}$$

To illustrate the consequences, the potential offspring of one H atom in an O_2/H_2 gas mixture are demonstrated in Figure 7.1. This shows that after five generations the number of chain carriers may have grown to 13. Potentially, it needs only the dissociation of one molecule of H_2 to set off a reaction whose rate will increase exponentially.

Experimentally, when a mixture of H_2 and O_2, in the stoichiometric proportions of 2:1, is admitted to a pre-heated vessel, an explosion does not necessarily ensue. As illustrated in Figure 7.2, it depends on the temperature and the pressure. While the general shape of the explosion boundary varies little, the fine details of the curve are a function of the size and the surface material of the reaction vessel.

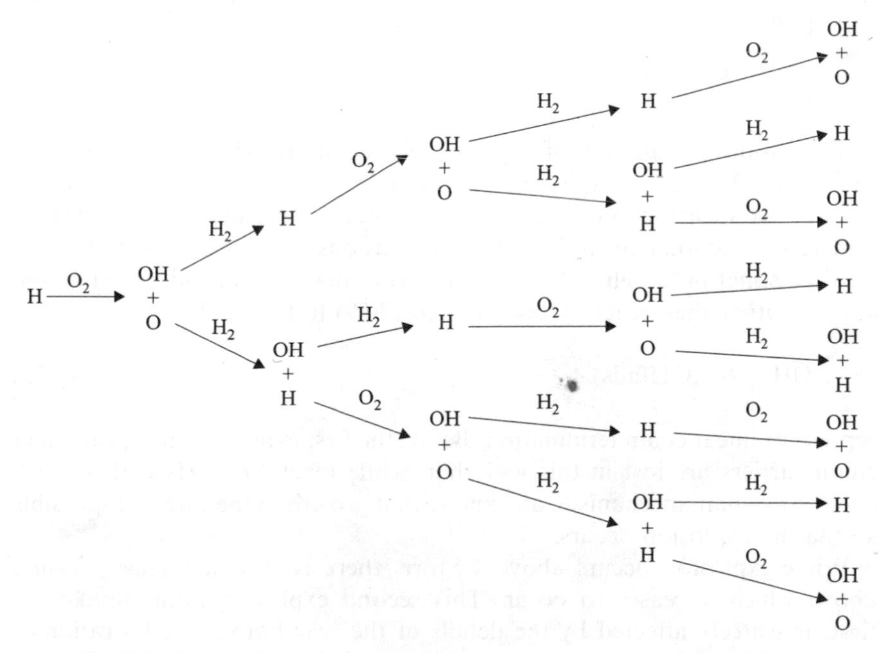

Fig. 7.1 A branching chain reaction.

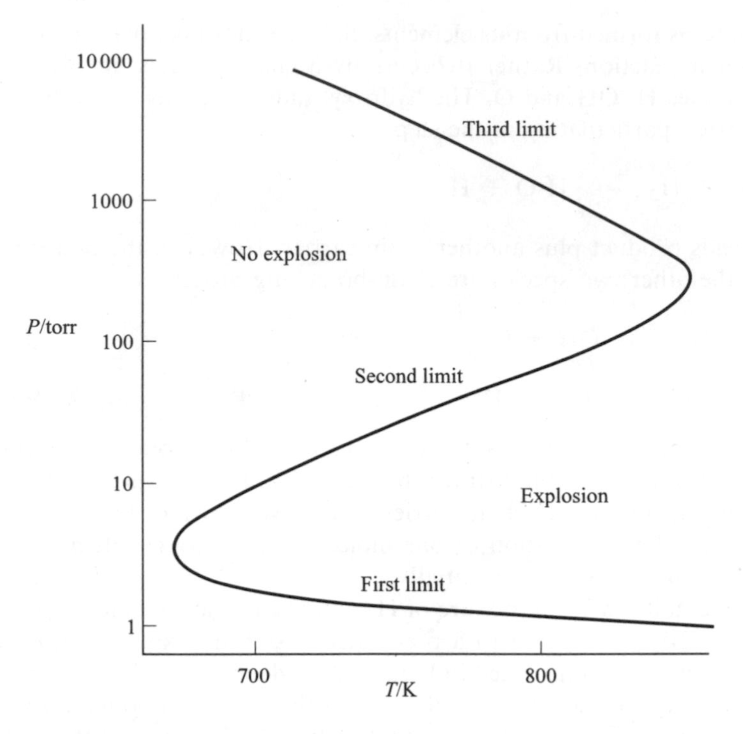

Fig. 7.2 The explosion boundary for stoichiometric mixtures of H_2 and O_2 in a KCl-coated vessel of diameter 7.4 cm. (Data from Lewis, B. and G. von Elbe, 1953, *Combustion, Flames and Explosion of Gases*, New York, Academic, p. 29.)

The outline explanation of Figure 7.2 is readily provided. At a temperature of *ca* 773 K, no explosion occurs unless the total pressure exceeds 1.5 torr. At such low pressures, the distance a radical may travel before it undergoes reaction may approach the dimensions of the vessel. Thus there is now a distinct probability that a chain carrier may become adsorbed on the surface rather than react by one of steps (7.69) to (7.71). The process

$$OH \rightarrow OH(ads) \tag{7.72}$$

represents linear chain termination. Below the first explosion limit, sufficient chain carriers are lost in this way that, while reaction of H_2 with O_2 will occur by a chain mechanism, an exponential growth in the rate is impossible so that no explosion occurs.

While explosion occurs above 1.5 torr, there is then a higher pressure above which it ceases to occur. This second explosion limit, unlike the first, is scarcely affected by the details of the reaction vessel. Its rationale is also quite different.

Reaction (7.71) involves an atom and a diatomic molecule. There are two possible outcomes of such a process, namely, the formation of the two radical species written in equation (7.71) or the formation of the triatomic radical, HO_2. The difficulty of forming HO_2 is analogous to that of forming the I_2 molecule from the reaction of I + I, discussed in Section 3.4. Thus the probability of forming the HO_2 radical is improved by the action of a third body, i.e.

$$H + O_2 + M \rightarrow HO_2 + M \tag{7.73}$$

However, the HO_2 radical is fairly unreactive in this system, so that it mostly becomes adsorbed on the vessel surface. In effect, this reaction represents linear chain termination, and above a certain pressure, where this termolecular reaction has become increasingly probable at the expense of the bimolecular (7.70), chain branching will have given way to chain termination to such an extent that no explosion can occur.

The pressure dependence of the second limit may be treated quantitatively, on the assumption that it represents some constant ratio, f, of the rates of reactions (7.70) and (7.73), i.e.

$$k_b[H][O_2] = fk_t[H][O_2][M] \tag{7.74}$$

where k_b and k_t are the rate constants of reactions (7.70) and (7.73). Thus

$$\begin{aligned}[M] &= k_b/fk_t \\ &= \frac{A_b}{fA_t}\frac{e^{-E_b/RT}}{e^{-E_t/RT}}\end{aligned} \tag{7.75}$$

Taking natural logs of both sides we have:

$$\ln[M] = \ln\left(\frac{A_b}{fA_t}\right) - \frac{E_b - E_t}{RT} \tag{7.76}$$

As we have seen in Chapter 3, the Arrhenius activation energy of a reaction involving a third body may be slightly negative: E_b will be positive, so the plot of ln[M] against T^{-1} will have a negative slope. Thus as the pressure is increased, one expects the temperature of the second explosion limit to rise, as it does in Figure 7.2.

There are two putative origins of the third explosion limit, which at 773 K occurs at *ca* 3000 torr. One explanation is that at higher pressures, where the HO_2 radical takes much longer to diffuse to the walls, it may well engage in the gas phase reaction:

$$HO_2 + H_2 \rightarrow H_2O_2 + H \tag{7.77}$$

This means that at higher pressures a sharply increasing proportion of the HO_2 radicals are capable of acting as chain carriers. Reaction (7.77) occurs the more readily the higher the temperature, which is consistent with the effect of temperature on the third explosion limit.

The alternative explanation, widely believed to be the operative one, is that the third limit is a thermal explosion. Reaction (7.68) is highly exothermic, and at 773 K, the non-explosive reaction occurs quite rapidly, releasing a large amount of energy. As the gas pressure is increased, the amount of energy released per unit volume rises in proportion, but the coefficient of thermal conductivity does not. Thus the temperature in the interior of the reaction vessel rises to a greater extent the higher the pressure. This rise causes an acceleration in the reaction, which causes the temperature to rise even more. Unless the supply of H_2 and O_2 is exhausted first, an explosion is the inevitable result. The higher the reaction vessel temperature, the lower is the pressure at which the explosion will occur.

7.6.2 Nuclear chain fission

The target of the US Manhattan Project, 1942–45, provides another example of a branching chain reaction, namely the induced chain fission of a large atomic nucleus. The significant process here is that in which a slow neutron is captured by a very large nucleus, causing the latter to split into two major fragments. The favoured ratio of neutrons to protons is of course higher for very large than for medium to medium large nuclides, so there will be several neutrons surplus to the requirements of these fragments.

The fission process does not yield unique products and many of the nuclides produced are more neutron-rich than any of the stable isotopes of that element. An attempt to indicate characteristic products is shown in equation (7.78):

$$ {}^{1}_{0}\mathrm{n} + {}^{235}_{92}\mathrm{U} \quad \rightarrow \quad {}^{94}_{38}\mathrm{Sr} + {}^{139}_{54}\mathrm{Xe} + 3\,{}^{1}_{0}\mathrm{n} \qquad (7.78) $$

This shows three neutrons produced from this fission step: the average is 2.5, so that this nuclear chain reaction branches rather sharply. It also releases a huge amount of energy: reaction (7.78) would correspond to the release of *ca* 200 MeV.

To achieve chain fission, there must be sufficient ^{235}U so that not too large a proportion of the neutrons is "lost". The rate of loss must depend on the geometry of the sample: the most efficient arrangement is obviously spherical. The minimum amount of ^{235}U has been given the name the "critical mass", whose value deserves to be one of the unpublished secrets of nuclear weapons technology.

The term "critical mass" is widely used in many a context, although sometimes the analogy with nuclear chain fission may be extremely tenuous. However, the behaviour of an H_2/O_2 mixture at a pressure below the first explosion limit shows considerable resemblance to that of a sub-critical amount of ^{235}U. At the first explosion limit a greater amount of H_2/O_2 is present, which might be thought of as the "critical mass" under the prevailing experimental conditions.

Suggested reading

Gimblett, F. G. R., 1970. *Introduction to the Kinetics of Chemical Chain Reactions*, London: McGraw-Hill.

Dainton, F. S., 1966. *Chain Reactions: An Introduction*, 2nd edition. London: Methuen.

Bamford, C. H. and C. F. H. Tipper (editors), 1976. *Comprehensive Chemical Kinetics*, Volumes 14A and 15. Amsterdam: Elsevier.

Lewis, B. and G. von Elbe, 1961. *Combustion, Flames and Explosions of Gases*, 2nd edition. New York: Academic Press.

Baldwin, R. R. and R. W. Walker, 1972. Branching-chain reactions: the hydrogen–oxygen reaction. In *Essays in Chemistry* (eds Bradley, J. N., R. D. Gillard and R. F. Hudson), Vol. 3, pp. 1–37. London: Academic Press.

Problems

7.1 A reaction proceeds by the following chain mechanism:

$$\begin{aligned} I &\xrightarrow{1} R^{\cdot} + R^{\cdot} \\ R^{\cdot} + M &\xrightarrow{2} P + R'^{\cdot} \\ R'^{\cdot} + N &\xrightarrow{3} Q + R^{\cdot} \\ R^{\cdot} + R^{\cdot} &\xrightarrow{4} Z \end{aligned}$$

(a) Write the stoichiometric equation for the chemical reaction that is taking place.

(b) Derive a rate expression for the reaction in terms of the concentrations of I, M and N and the rate constants of the four reaction steps.

7.2 The Rice–Herzfeld mechanism for the pyrolysis of ethane is as follows:

$$\begin{aligned} C_2H_6 &\xrightarrow{1} CH_3^{\cdot} + CH_3^{\cdot} \\ CH_3^{\cdot} + C_2H_6 &\xrightarrow{2} CH_4 + C_2H_5^{\cdot} \\ C_2H_5^{\cdot} &\xrightarrow{3} C_2H_4 + H^{\cdot} \\ H^{\cdot} + C_2H_6 &\xrightarrow{4} H_2 + C_2H_5^{\cdot} \\ H^{\cdot} + C_2H_5^{\cdot} &\xrightarrow{5} C_2H_6 \end{aligned}$$

(a) Find a rate expression for the formation of ethene in terms of the concentration of ethane and the rate constants of these five reaction steps.

(b) On the basis of the above mechanism, if a mixture of C_2H_6 and C_2D_6 were pyrolysed, would you expect (i) H_2 and D_2 only or (ii) HD only or (iii) H_2, HD and D_2 to be formed?

7.3 In a Teflon-lined reaction vessel, the reaction

$$F_2 + CCl_3F = ClF + CCl_2F_2$$

was found, at around 550 K, to be first order in CCl_3F and one-half order in F_2. Suggest a mechanism and show that it would lead to a rate expression consistent with these requirements. (Data from Foon, R. and K. B. Tait, *J. Chem. Soc. Faraday Trans I*, 1972, **68**, 104–11.)

7.4 For the reaction of H_2 with Br_2, the activation energies of the propagation steps,

$$Br^{\cdot} + H_2 \xrightarrow{2} HBr + H^{\cdot}$$

$$H^{\cdot} + Br_2 \xrightarrow{3} HBr + Br^{\cdot}$$

have been determined as 76 and 4 kJ mol^{-1}, respectively.

(a) In a stoichiometric mixture of H_2 and Br_2 at 550 K, estimate by what factor the steady state concentration of $Br^{\cdot}$ will exceed that of $H^{\cdot}$.

(b) How do these steady state concentrations of $Br^{\cdot}$ and $H^{\cdot}$ compare with that of $Br^{\cdot}$ expected at the same temperature in a vessel containing Br_2 alone at the same partial pressure and with that of $H^{\cdot}$ expected at the same temperature in a vessel containing H_2 alone at the same partial pressure?

7.5 For the gas phase reaction,

$$H_2 + NO_2 = H_2O + NO$$

the following radical chain mechanism has been proposed:

$$H_2 + NO_2 \xrightarrow{2} H^{\cdot} + HONO$$

$$H^{\cdot} + NO_2 \xrightarrow{2} OH^{\cdot} + NO$$

$$OH^{\cdot} + H_2 \xrightarrow{3} H_2O + H^{\cdot}$$

$$OH^{\cdot} + NO_2 \xrightarrow{4} HNO_3$$

Show that this mechanism predicts a rate equation of the form,

$$\frac{d[H_2O]}{dt} = k'[H_2]^2$$

and explain why it does not involve the concentration of NO_2, which is involved in the initiation step.

7.6 Hexamethyldisilane (A), $Me_3SiSiMe_3$, isomerises to trimethylsilyl(dimethylsilyl) methane, B. This conversion occurs the more readily in the presence of toluene vapour and the following mechanism for this reaction, at the temperatures in the range 700–800 K, has been proposed:

$$Me_3SiSiMe_3 \xrightarrow{1} 2Me_3Si^{\cdot}$$

$$Me_3Si^{\cdot} + PhMe \xrightarrow{2} Me_3SiH + Ph\dot{C}H_2$$

$$Ph\dot{C}H_2 + Me_3SiSiMe_3 \underset{-3}{\overset{3}{\rightleftarrows}} PhMe + Me_3SiSi(\dot{C}H_2)Me_2$$

$$Me_3SiSi(\dot{C}H_2)Me_2 \xrightarrow{4} Me_3SiCH_2\dot{S}iMe_2$$

$$Me_3SiCH_2\dot{S}iMe_2 + PhMe \xrightarrow{5} Me_3SiCH_2Si(H)Me_2 + Ph\dot{C}H_2$$

$$Ph\dot{C}H_2 + Ph\dot{C}H_2 \xrightarrow{6} PhCH_2CH_2Ph$$

(a) On the basis of this mechanism, work out the rate expression for the formation of B in terms of the concentrations of A and of toluene, and of the relevant rate constants.

(b) Under the conditions used in a series of experiments, $k_4 \gg k_{-3}$ [PhMe]. Evaluate the Arrhenius activation energy for forming (B) in terms of the activation energies of the individual steps of the reaction. (From Davidson, I. M. T., C. Eaborn and J. M. Simmie, *J. Chem. Soc. Faraday Trans I*, 1974, **70**, 249–52.)

8 Heterogeneous catalysis

The role of a catalyst in a different (usually solid) phase in bringing about a reaction of species in the gaseous or solution phases is a stimulating one. For this phenomenon to occur, at least one of the reacting species must undergo chemisorption on the catalyst surface, but for ease of reaction, it should not be too strongly adsorbed. The manner in which the rate of the catalysed reaction depends on the concentration of the reactant (and perhaps the product) species is very much a consequence of the way in which the concentrations of the various adsorbed entities involved in the reaction are affected by these concentrations.

8.1 Introduction

There are some chemical reactions of which the rate is caused to increase when another substance, not consumed in the reaction, is added to the containing vessel. For this phenomenon, Berzelius, in 1835, coined the name "catalysis". In some instances, the effect of the catalyst is merely to cause a slight acceleration, so that careful measurements are necessary to detect this effect. In other cases, there may be no observable chemical reaction in the absence of the catalyst but a substantial and unmistakeable one in its presence. This is particularly true of some gas phase reactions for which specific classes of solids act as catalysts, of the heterogeneous type.

Efforts to understand the operation of heterogeneous catalysis showed that the reaction must be taking place on the catalyst surface, and that the process of adsorption was fundamental to this understanding. It is then logical to start the systematic development by looking at the interaction of potential reactants with the catalyst surface.

8.2 Chemisorption and the chemisorbed state

The term adsorption is used to denote the attachment of one substance to the surface of another. We shall discuss it in terms of the adsorption of a gas on the surface of a solid, but the phenomenon also occurs with other combinations of phases.

The sub-division of adsorption into physical and chemical aspects is notional rather than absolute. The conception is that the forces holding the adsorbed molecule (the "adsorbate") to the surface of the solid (the "adsorbent") may be analogous either to the van der Waals attractive forces operative between molecules within a liquid or to a chemical bond between atoms within a molecule. The former is called *physisorption* and the latter *chemisorption*.

A physisorption process may be likened to the liquefaction of a vapour and so it will occur rapidly. Chemisorption is a class of chemical reaction and so it may be slow. However, it will be limited to a single layer of adsorbate on the surface. On the other hand, physisorption may well entail additional adsorption on top of the first adsorbed layer, since the forces between the adsorbate and the surface are similar to those between successive layers of adsorbate. The amount of physically adsorbed adsorbate is thus potentially unlimited.

The standard enthalpy changes involved for physisorption and chemisorption typically fall into different ranges. Physisorption is usually exothermic, with $\Delta H^{\ominus}$ around –20 to –50 kJ mol^{-1}; for chemisorption, the possible range of values is much wider, as for any chemical reaction. For example, for the chemisorption of O_2 on molybdenum, $\Delta H^{\ominus}$ is –720 kJ mol^{-1}, but for the chemisorption of $-H_2$ on porous glass it is *ca* +30 kJ mol^{-1}. Whereas a physisorption process is always easily reversible, chemisorption is less so. In some cases, such as the very strong chemisorption of O_2 on certain transition metals, the direct desorption process may be impossible to achieve.

There are basically two routes for the measurement of heats of adsorption. One is calorimetric and requires that the adsorbent has been made to form part of an adiabatic system whose temperature rise, consequent on the adsorption of an aliquot of adsorbate, can be measured. The relevant calculation then yields the mean differential heat of adsorption for this aliquot of adsorbate. From adding other aliquots, it may be determined how the differential heat of adsorption varies with the extent of coverage, θ, of the surface of the adsorbent.

The alternative approach is a careful measurement of the amount of adsorbate in equilibrium with a known adsorbate pressure, that is, of the adsorption isotherm, at several different temperatures. The pressure required to achieve a particular value of θ varies with the temperature in a manner reminiscent of the Clausius–Clapeyron equation for the vaporisation of a liquid. We have,

$$\ln p = -\frac{\Delta H_\theta}{RT} + \text{const} \tag{8.1}$$

where the enthalpy of adsorption ΔH_θ is the isosteric value at that particular fraction of the surface covered. From the data, the isosteric heat of adsorption (which differs from the differential heat of adsorption only by RT) may be found at various values of θ.

Using either method, the gist of the findings is that the adsorption process becomes less exothermal the higher the value of θ. There are two simple ways in which to interpret such a result. It is conceivable that the adsorption sites are intrinsically heterogeneous, with some of them capable of interacting much more strongly than others with the adsorbate. Provided that there is any mobility of the adsorbed species, one then expects that, preferentially, the strongly adsorbing sites will be occupied first.

The alternative is that, while all adsorption sites are initially equivalent, as sites become occupied the capacity of neighbouring sites to adsorb strongly is reduced, whether on account of geometric or electronic considerations. This is the concept of induced heterogeneity.

8.3 The Langmuir adsorption isotherm

The adsorption of various gases on metal surfaces was studied in some detail by Langmuir, whose work won a Nobel Prize. He found that in several cases the amount adsorbed rose sharply in the initial stages and then tended to level out. He reasoned that the value towards which the amount adsorbed was rising must represent a limit of feasibility: the most likely scenario was that it corresponded to a surface monolayer with every site occupied. He also derived an equation based on a simple model of surface behaviour, inter-relating the surface coverage with the pressure. This is widely known as the Langmuir Adsorption Isotherm.

The model involved makes the following assumptions:

(i) The surface of the adsorbent consists of a finite number of equivalent adsorption sites, each capable of taking one adsorbed molecule;
(ii) Adsorption may occur when a gaseous molecule, A, collides with a vacant site, with the sticking coefficient independent of θ, the fraction of sites that are occupied.
(iii) The probability that an adsorbed molecule may desorb is also independent of θ.

The rate of adsorption will contain the factor $\exp(-E_{ads}/RT)$ and that of desorption the factor $\exp(-E_{des}/RT)$, where E_{ads} and E_{des} are the activation energies for adsorption and desorption on a surface site. The inter-relationship of these quantities is shown in Figure 8.1.

At equilibrium, the rates of adsorption and of desorption may be equated:

$$p_A(1-\theta)b_1 \exp(-E_{ads}/RT) = \theta b_2 \exp(-E_{des}/RT) \qquad (8.2)$$

where p_A denotes the pressure of A and b_1 and b_2 are appropriate constants. Rearranging, using the fact that $\Delta H_{ads} = E_{ads} - E_{des}$, and putting $(b_1/b_2)\exp(-\Delta H_{ads}/RT) = a$, we have:

$$\theta = \frac{ap_A}{1+ap_A} \qquad (8.3)$$

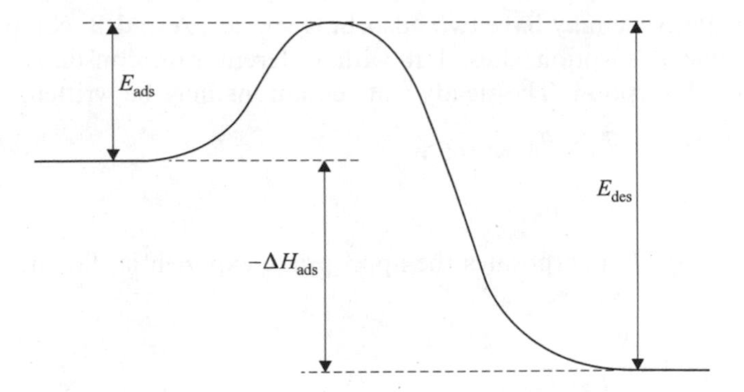

Fig. 8.1 Enthalpy profile for the adsorption process showing the inter-relationship of the activation energy for adsorption, E_{ads}, the enthalpy of adsorption, ΔH_{ads}, and the activation energy for desorption, E_{des}.

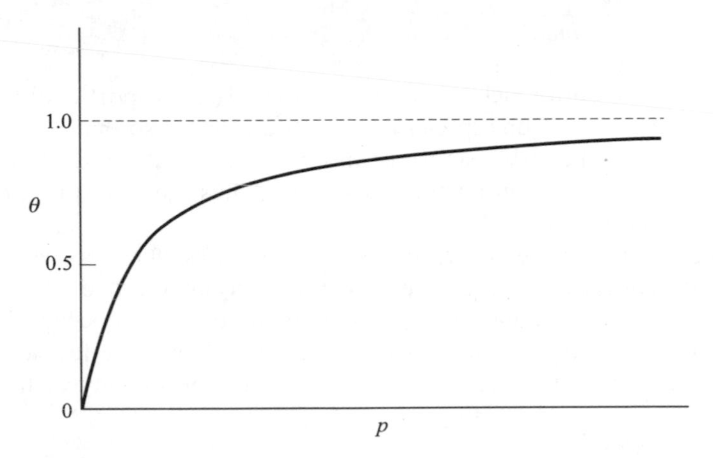

Fig. 8.2 The Langmuir isotherm for monolayer adsorption. The dependence of θ, the fractional surface coverage, on p, the pressure of the adsorbate.

This is the Langmuir isotherm, whose implications are illustrated on Figure 8.2.

For very small values of p_A, where $ap_A \ll 1$, it follows that $\theta \approx ap_A$ and so is proportional to p_A. But at large or very large values of p_A where $ap_A \gg 1$, then θ is approximately equal to unity: that is, the surface has virtually a monolayer of A adsorbed on it. Taking the reciprocal of each side of equation (8.3) leads to the expression:

$$\frac{1}{\theta} = 1 + \frac{1}{ap_A} \tag{8.4}$$

Thus the reciprocal of θ should be a linear function of the reciprocal of the pressure.

In some instances we may have two adsorbing species, A and B, competing for the same adsorption sites, but with different rate constants for adsorption and desorption. The steady state equations may be written,

$$k_1\theta_A = k_2 p_A(1 - \theta_A - \theta_B) \tag{8.5}$$

$$k_3\theta_B = k_4 p_B(1 - \theta_A - \theta_B) \tag{8.6}$$

where each constant k incorporates the appropriate exponential factor.

$$\theta_A = \frac{\dfrac{k_1}{k_2} p_A}{1 + \dfrac{k_1}{k_2} p_A + \dfrac{k_3}{k_4} p_B} = \frac{a p_A}{1 + a p_A + b p_B} \tag{8.7}$$

and $$\theta_B = \frac{b p_B}{1 + a p_A + b p_B} \tag{8.8}$$

If ap_A and bp_B are both much less than one, then θ_A is proportional to p_A and θ_B to p_B. However, if one species is strongly adsorbed, so that $ap_A \gg 1$, the consequence is that whereas θ_A is approximately unity and therefore essentially constant, θ_B is proportional to p_B/p_A. Thus the adsorption of B is being inhibited by that of A.

Another variation of the adsorption conditions arises where, as with H_2, the adsorbate undergoes dissociative adsorption, occupying two sites. For adsorption to occur, two adjacent empty sites are needed and if occupancy is random, the probability of this is given by $(1 - \theta_A)^2$. Likewise for desorption, two adjacent occupied sites are needed and the probability of that is given by $\theta_A{}^2$.

Thus we have:

$$k\theta_A^2 = k' p_A(1 - \theta_A)^2 \tag{8.9}$$

By taking the square root of each side, this equation comes to resemble equation (8.2) and leads to:

$$\theta_A = \frac{a' p_A^{1/2}}{1 + a' p_A^{1/2}} \tag{8.10}$$

This shows that in this situation the fractional surface coverage rises towards one in a slightly different way from that described in equation (8.3). In the early stages, θ_A is proportional to the square root of p_A. Also, $\theta_A{}^{-1}$ is now a linear function of $p_A{}^{-1/2}$, and the constant a' contains the factor $\exp(-\Delta H_{ads}/2RT)$.

It should be pointed out that any version of the Langmuir adsorption isotherm is derived on the assumption that the heat of adsorption is invariant. Experimental evidence shows that this is scarcely ever the case and so the basis of the isotherm is suspect. It is therefore prudent to apply the Langmuir adsorption isotherm with some caution.

8.4 Reactions on a catalyst surface

Solid catalysts may be used to accelerate reactions involving species either in the gas phase or in solution. The major difference between these is that in the latter case it is much more likely that the rate of the process will be controlled by diffusion, either of the reactants to the catalyst or of the products from the catalyst. The considerations involved have been discussed in Section 2.1. In order to focus on processes involving the catalyst, the present discussion will be confined to reactions involving only the gas phase.

Let us start by considering a catalysed reaction of the simplest conceivable mechanism. A species A is chemisorbed on the surface of the catalyst. When it reacts, in the same step it desorbs from the surface as the product, B. The reaction scheme is thus:

$$\left.\begin{array}{rcl} \mathrm{A(g)} & \rightleftharpoons & \mathrm{A(ads)} \\ \mathrm{A(ads)} & \xrightarrow{r} & \mathrm{B(g)} \end{array}\right\} \qquad (8.11)$$

The rate of formation of B will be given by the product of θ_A and the relevant rate constant, k_r, for the reaction step. Using the Langmuir isotherm for θ_A, we have:

$$\begin{aligned} \text{Rate} &= k_r\theta_A \\ &= \frac{k_r a p_A}{1 + a p_A} \end{aligned} \qquad (8.12)$$

As was remarked previously, there are two extreme regions of the Langmuir isotherm, namely where $ap_A \ll 1$, which corresponds to weak adsorption, and where $ap_A \gg 1$. For weak adsorption, we have,

$$(\text{Rate})_{\text{w.a.}} = k_r a p_A \qquad (8.12a)$$

and for strong adsorption,

$$(\text{Rate})_{\text{s.a.}} = k_r \qquad (8.12b)$$

This indicates that the reaction may display either first order (equation 8.12a) or zero order kinetics (equation 8.12.b), depending on the strength of adsorption of the reactant. These equations reflect the two extremes, so any exponent between zero and one must be possible. For weak adsorption, the Arrhenius activation energy of the catalysed reaction is given by:

$$(E_a)_{\text{w.a.}} = -R\,\frac{\mathrm{d}\ln(k_r a)}{\mathrm{d}(1/T)} \qquad (8.13)$$

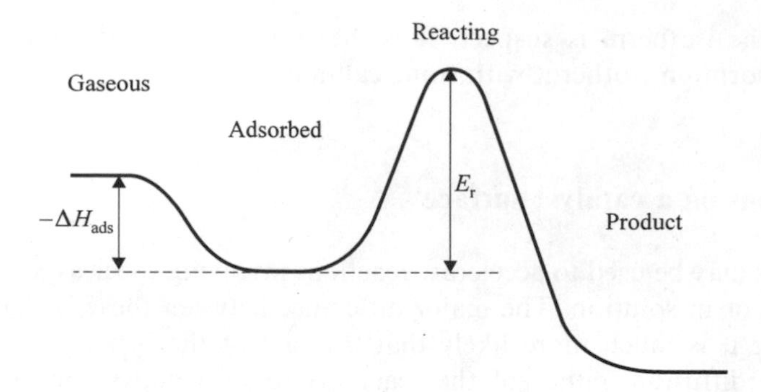

Fig. 8.3 Enthalpy profile for reaction scheme (8.11), where ΔH_{ads}, refers to the adsorption of A and E_r to the reaction of adsorbed A to give B.

Assuming that the rate constant k_r may be equated to $A_r\exp(-E_r/RT)$ and remembering that the constant a contains the factor, $\exp(-\Delta H_{ads}/RT)$, this gives:

$$(E_a)_{w.a.} = E_r + \Delta H_{ads} \tag{8.13a}$$

$$(E_a)_{s.a.} = E_r \tag{8.13b}$$

Since ΔH_{ads} is a negative quantity, this activation energy is greater than in the case of weak adsorption. The energy profile corresponding to scheme (8.11) is sketched in Figure 8.3, which shows that where A is strongly adsorbed or p_A is sufficiently high that the surface is almost covered, the effective energy barrier to reaction is E_r. However, where A is only weakly adsorbed so that the surface is sparsely covered, the effective barrier is less than this.

Quite often, two reacting species are involved in a catalysed reaction. An obvious reaction scheme is that each undergoes adsorption and that the two adsorbed species then undergo reaction to yield the product.

$$\left.\begin{array}{rcl} A(g) & \rightleftharpoons & A(ads) \\ B(g) & \rightleftharpoons & B(ads) \\ A(ads) + B(ads) & \xrightarrow{r} & P(g) \end{array}\right\} \tag{8.14}$$

This is known as a Langmuir–Hinshelwood mechanism and the reaction rate is easily expressed in terms of the fractional surface coverages of A and B, and, using equations (8.7) and (8.8), in terms of the partial pressures of A and B:

$$\begin{aligned} \text{Rate} &= k_r\theta_A\theta_B \\ &= k_r\frac{ap_A bp_B}{(1 + ap_A + bp_B)^2} \end{aligned} \tag{8.15}$$

If both A and B are weakly adsorbed, then the rate will be proportional to $p_A p_B$ and the reaction will exhibit second order kinetics. However, if B is strongly adsorbed and A weakly, the rate will be proportional to $p_A p_B^{-1}$. Under these conditions, the strongly adsorbed B is inhibiting the adsorption of A which is essential for reaction. More probably, both θ_A and θ_B will have moderate values and the reaction orders with respect to both reactants will have non-integral values between 1 and −1, with the proviso that no more than one of these exponents should be negative.

As an alternative to the Langmuir–Hinshelwood mechanism, Eley and Rideal proposed that reaction might occur directly between one species in the adsorbed state and the other in the gas phase. The scheme is thus:

$$\left.\begin{array}{rcl} A(g) & \rightleftharpoons & A(ads) \\ A(ads) + B(g) & \xrightarrow{k_r} & P(g) \end{array}\right\} \qquad (8.16)$$

In general, B may be expected to adsorb, even though no role has been proposed for the adsorbed B. Thus we have:

$$\begin{aligned} \text{Rate} &= k_r \theta_A p_B \\ &= k_r \frac{a p_A p_B}{(1 + a p_A + b p_B)} \end{aligned} \qquad (8.17)$$

The rate expressions arising from the Langmuir–Hinshelwood and the Eley–Rideal mechanisms, equations (8.15) and (8.17), differ slightly but significantly. Whereas equation (8.17) implies that, if p_B is kept constant and p_A is increased, then the rate should rise monotonically, equation (8.15) predicts that the rate should attain a maximum and then decline. This maximum may be shown to occur at $p_A = (1 + b p_B)/a$: at higher pressures of A, the rate declines slowly, falling by only 11% when p_A is twice the optimum value.

On occasions, the presence or absence of a maximum in the rate as a function of the pressure of either reactant, keeping constant the pressure of the other, has been used as an argument for one or other of these mechanisms. Bearing in mind the qualifications above regarding the total reliability of the Langmuir isotherm, such evidence would seem less than conclusive.

8.5 Catalytic exchange of alkanes with deuterium

Soon after deuterium's discovery in 1933, it became available for use as a labelled form of hydrogen and was applied in catalytic studies. From the ease with which a D atom could be incorporated into a molecule it soon became clear that, on a catalyst surface, a C–H bond might be far more labile than a C–C bond, even though the latter has the lower bond dissociation energy.

A study of deuterium exchange reactions thus sheds light on the mechanism and the species involved in these processes. In regard to adsorbed species it has the merit, not universally enjoyed by spectroscopic techniques, that the information derived relates entirely to species taking part in surface reactions, as distinct from those which may be too strongly adsorbed so to do.

A technique used extensively in the study of catalytic exchange reactions was pioneered by Kemball, around 1950. A film of a transition metal was formed *in vacuo* by evaporation from an appropriate filament, to furnish the polycrystalline, high surface area catalyst for the reaction. A mixture of the alkane and deuterium was then admitted to the catalyst, at or below the lowest temperature at which a slow reaction was obtained. A fine capillary attached to the reaction vessel allowed a few percent per hour of the reaction mixture to leak away into the ionisation chamber of a mass spectrometer, providing sufficient alkane to permit its isotopic analysis but without undue disturbance to the reaction at the catalyst.

In the analysis of the partly exchanged alkanes, an advantage of using the mass spectrometer is that it can yield not merely the amount of deuterium incorporated, but its distribution. However, in a primary amine, for example, the distribution of the number of D atoms in the amine group can have no mechanistic import in view of the occurrence of the reaction,

$$RNH_2 + RND_2 \rightleftharpoons 2RNHD \qquad (8.18)$$

which takes place very readily, presumably on the slightly acidic surface of the glass vessel. The distribution of the number of D atoms in a partly exchanged alkane is significant only because alkanes do not engage in such equilibration processes. It may be remarked that the information acquired in this way is not accessible using radioisotopes.

Electron impact on an alkane molecule of relative molar mass M, will generate, among others, the parent ion M^+. To monitor deuterium exchange, it is desirable to achieve conditions which minimise the size of the fragment ions at $(M - 1)^+$ and $(M - 2)^+$ by loss of one or of two H atoms from the molecule. By eliminating these fragment ions, then the growth of the $(M + 1)^+$ peaks above the values due to naturally occurring ^{13}C is a measure of the amounts of the d_1- and d_2-alkanes present. Kemball achieved this by employing very low electron accelerating voltages, just above the appearance potential of the parent ion.

It is sometimes difficult to obtain information on the distribution of D atoms within a molecule: for example, to tell whether d_2-ethane is mostly CH_3CHD_2 or CH_2DCH_2D. In some cases, fragment ions may assist appreciably in this quest.

In studying the reaction of ethane with D_2, it was found that over most transition metals the initial yield of several products was finite. This ruled out the possibility that exchange takes place only one atom at a time per visit to the catalyst surface. In that case, d_1-ethane would be the only initial product and the initial rates of formation of d_2-ethane, d_3-ethane, etc. would all be zero (see Section 3.2). On some transition metals, such as molybdenum, in the initial distribution it was seen that $Y_1 > Y_2 > Y_3 > Y_4$, where Y_n

denotes the yield of d_n-ethane. On other metals, such as palladium, the order of the yields was reversed.

To explain these findings, Kemball proposed that there were two adsorbed intermediates formed in successive steps by the dissociative adsorption of ethane on the catalyst surface:

$$\left.\begin{aligned} C_2H_6(g) &\underset{-1}{\overset{1}{\rightleftarrows}} C_2H_5(ads) + H(ads) \\ C_2H_5(ads) &\underset{-2}{\overset{2}{\rightleftarrows}} C_2H_4(ads) + H(ads) \end{aligned}\right\} \qquad (8.19)$$

Since D_2 will also undergo dissociative chemisorption,

$$D_2(g) \rightleftharpoons 2D(ads) \qquad (8.20)$$

it is obvious that the intermediate C_2H_4(ads) may, when it undergoes step −2, pick up a D rather than an H atom, and that a similar result may ensue when C_2H_5(ads) undergoes step −1. If a highly exchanged ethane molecule is to be obtained, the behaviour of C_2H_5(ads) is crucial: it must have high tendency to engage in step 2 rather than in step −1.

To treat this question quantitatively, Kemball defined a parameter P:

$$P = \frac{\text{probability of step 2}}{\text{probability of step } -1} \qquad (8.21)$$

He calculated the distributions of deuteroethanes for various values of P and showed that, for $P = 0.25$, the figures were in acceptable agreement with the yields found from the molybdenum catalyst, as is shown in Table 8.1. The results on palladium were found to be consistent with the distribution calculated from $P = 18$.

On a platinum catalyst, the distribution showed twin maxima at d_1 and d_6. No single value of P could reproduce such a distribution, but it was found that a 50 : 50 combination of $P = 2.0$ and $P = 18$ gave acceptable agreement. This would imply that, for adsorbed ethyl radicals, a platinum surface has two different kinds of adsorption site.

Further studies of the deuterium exchange reactions of hydrocarbons showed that a simple extension of the above approach was not in general capable of explaining the results. For example, 2,2-dimethylpropane shows negligible multiple exchange on any metal, even though it seems feasible to form from it the mono-adsorbed and di-adsorbed radicals,

```
       CH3                         CH3   CH3
        |                             \ /
CH3 --- C --- CH3                      C
        |                             / \
       CH2                         CH2   CH2
        |                           |     |
        *          (I)              *     *     (II)
```

Table 8.1 Observed and calculated initial distributions of products for the exchange of ethane on films.

	Percentages of isotopic species					
Catalyst	C_2H_5D	$C_2H_4D_2$	$C_2H_3D_3$	$C_2H_2D_4$	C_2HD_5	C_2D_6
Mo obs.	82	14	3.0	0.7	–	–
calc. $P = 0.25$	80.0	17.2	2.5	0.3	–	–
Pd obs.	5	6	8	11	19	51
calc. $P = 18$	5.3	7.3	8.7	11.2	16.9	50.6
Pt obs.	20	17	12	10	15	26
calc. $P = 2.0$ (50%)	16.6	14.3	9.5	5.7	2.9	1.0
calc. $P = 18$ (50%)	2.6	3.6	4.4	5.6	8.4	25.4
Total calc.	19.2	17.9	13.9	11.3	11.3	26.4

which should be capable of behaving in a similar manner to the two surface species proposed in scheme (8.19) in regard to the adsorption of ethane. However, with *n*-alkanes and cycloalkanes, very extensive multiple exchange occurs, but to a pattern not readily reproduced by an extension of the model so successful with ethane.

It has since been realised that the species C_2H_4(ads), used in scheme (8.19), is not a σ-bonded, α, β-diadsorbed species,

$$\begin{array}{ccc} CH_2 & — & CH_2 \\ | & & | \\ * & & * \end{array} \quad \text{(III)}$$

but rather a π-bonded entity, adsorbed on a single site of the catalyst,

$$CH_2 \overset{}{\underset{*}{\ddagger}} CH_2 \quad \text{(IV)}$$

and analogous to the behaviour of ethene as a ligand in transition metal complexes. If one also accepts the existence of π-allylic species,

$$R\diagdown CH \cdots CH \cdots CH \diagup R \text{ (π-allyl bonded to } * \text{)} \quad \text{(V)}$$

then the sequence, π-olefin complex, π-allyl complex, π-olefin complex would provide a route by which substantial deuterium exchange of a large molecule might be achieved on a single visit to the catalyst surface.

8.6 The catalytic oxidation of carbon monoxide

The oxidation of CO to CO_2 is a reaction which seems intrinsically simpler than many catalytic processes since it involves only two reactant molecules, both of them diatomic, and the product molecule is not chemisorbed on the catalyst surface. These circumstances seem more auspicious for using the kinetics of the reaction to shed light on its mechanism.

The reaction is catalysed by a number of Group VIII transition metals, of which palladium and platinum are perhaps the most efficient. In laboratory studies, a metal catalyst may be used in any one of a number of physical states, such as powder, supported particles, wire, foil, film and single crystal. Also, the reaction may be studied using any part of a huge range of accessible pressures. Thus it is scarcely to be expected that every study will reveal the identical dependence on reactant concentrations.

Early studies of the reaction yielded the result that, on both palladium and platinum, the rate is proportional to the pressure of O_2 and inversely proportional to the pressure of CO. This suggests that oxygen is weakly adsorbed and CO strongly adsorbed and that a Langmuir–Hinshelwood mechanism is operative. The full rate equation would be:

$$\begin{aligned}\text{Rate} &= k\theta_{\text{O}} \cdot \theta_{\text{CO}} \\ &= k \cdot \frac{ap_{\text{O}_2}}{1 + ap_{\text{O}_2} + bp_{\text{CO}}} \cdot \frac{bp_{\text{CO}}}{1 + ap_{\text{O}_2} + bp_{\text{CO}}}\end{aligned} \tag{8.22}$$

Carbon monoxide is of course much more strongly adsorbed on palladium than is oxygen, so that $\theta_{\text{CO}} \approx 1$. Also, the term bp_{CO} will be the largest one in each denominator, so that the rate equation may be simplified:

$$\text{Rate} \approx \frac{kap_{\text{O}_2}}{bp_{\text{CO}}} \tag{8.22a}$$

Thus the findings seem consistent with the Langmuir–Hinshelwood mechanism.

Sometimes in chemical kinetics the most persuasive evidence in favour of a mechanism is that which shows that the other options are precluded. Let us then consider the feasibility of each version of an Eley–Rideal mechanism:

$$\text{CO(ads)} + \text{O}_2\text{(g)} \rightarrow \text{CO}_2\text{(g)} + \text{O(ads)} \tag{X}$$

$$\text{O(ads)} + \text{CO(g)} \rightarrow \text{CO(g)} \tag{Y}$$

The first of these is known not to occur, in that if a palladium (or a platinum) surface saturated with CO is exposed to O_2, no CO_2 is immediately formed. A specially designed experiment, carried out at low pressures, showed that if a palladium surface covered with adsorbed oxygen is bombarded with CO, then CO_2 is formed only after a delay of a few seconds. The absence of any immediate formation of CO_2 indicates that it is not formed by the second, Eley–Rideal step, Y.

There is also positive evidence for the occurrence of the Langmuir–Hinshelwood mechanism. In a sophisticated experiment by Engel and Ertl, using the (111) plane of a palladium single crystal in an ultra-high vacuum system, a background pressure of 10^{-7} torr of O_2 gave a surface coverage of oxygen, θ_O, of at least 0.1. A molecular beam of CO, equivalent to the same pressure as that of O_2, was directed at the surface and a mass spectrometer was positioned to detect any CO_2 molecules emitted from the palladium surface before they could undergo collision with the walls.

Rapid modulation of the molecular beam of CO made it possible to determine the time lag between the opening of the shutter and the appearance of CO_2 at the detector. This time lag, which was many orders of magnitude greater than the sum of the transit times of the CO and CO_2 molecules involved, showed that the Eley–Rideal mechanism Y was not operative. The variation of the time lag with the temperature of the palladium surface confirmed that it originated in events on the surface and thus, the formation of CO_2 by a Langmuir–Hinshelwood mechanism.

8.7 Catalytic synthesis and decomposition of ammonia

The synthesis of ammonia,

$$N_2 + 3H_2 \quad = \quad 2NH_3 \tag{8.23}$$

is a reaction which readily illustrates that a catalyst causes an increase in the rates of both the forward and the reverse reactions. It is much easier, in the laboratory, to achieve conditions where the decomposition reaction is uncomplicated by the reverse process than to do the same for ammonia synthesis, so studies of catalytic ammonia decomposition are easier to perform, simpler to interpret kinetically and of equal commercial relevance.

Most studies of the reaction, using a transition metal catalyst such as iron at *ca* 700 K, have reported that the rate of the decomposition is proportional to the ammonia pressure to some positive index and to the hydrogen pressure to a negative power. (The presence of H_2 is essential, since otherwise ammonia will react with the metal to form a nitride.) This negative exponent implies that the reaction is inhibited by hydrogen and suggests that this reaction product is strongly adsorbed. However, this approach is over-simplistic.

Considering the issue carefully, the following facts are seen to be relevant:

(i) Ammonia readily undergoes catalytic exchange with deuterium at temperatures far below those required for decomposition.

(ii) Where it is thermodynamically feasible, ammonia readily reacts with iron to form the nitride, Fe_4N:

$$NH_3(g) + 4Fe(s) \rightleftharpoons Fe_4N(s) + \tfrac{3}{2}H_2(g) \tag{8.24}$$

Moreover, the hydrogenation of Fe_4N to NH_3 also occurs readily.

(iii) The decomposition of Fe_4N, in the absence of hydrogen, to iron and N_2 proceeds slowly even in the temperature range required for the catalytic decomposition of ammonia.

(iv) On an iron catalyst, the exchange of nitrogen isotopes,

$$^{14}N_2 + {}^{15}N_2 \rightleftharpoons 2\,{}^{14}N^{15}N \tag{8.25}$$

occurs very slowly, whereas H_2/D_2 exchange is rapid at low temperatures.

These observations support the contention that, on an iron catalyst, the stepwise removal of the H atoms from an adsorbed ammonia molecule occurs very readily, as do the adsorption and desorption of hydrogen. However, the adsorption and desorption of nitrogen at the iron catalyst do not occur readily and so these processes constitute the respective slow steps in the synthesis and decomposition reactions. Thus we have,

$$\begin{array}{l} NH_3(g) \rightleftarrows NH_3(ads) \rightleftarrows NH_2(ads) + H(ads) \\ \qquad\qquad\qquad\qquad\qquad \swarrow\!\!\nearrow \\ \qquad\qquad NH(ads) + 2H(ads) \rightleftarrows N(ads) + 3H(ads) \\ 2H(ads) \rightleftarrows H_2(g) \\ 2N(ads) \xrightarrow{slow} N_2(g) \end{array} \tag{8.26}$$

Since the rate of ammonia decomposition is very much an issue of the desorption of nitrogen, it is useful to approach the question of the degree of occupancy of the adsorption sites by N atoms by way of the pressure of nitrogen which *would* be in equilibrium with the actual concentrations of NH_3 and H_2 at the catalyst. If the equilibrium constant of reaction (8.23) is K, then this fictitious pressure, $p^*_{N_2}$, is given by:

$$p^*_{N_2} = \frac{1}{K}\frac{(p_{NH_3})^2}{(p_{H_2})^3} \tag{8.27}$$

The usefulness of this quantity is that, on the basis of scheme (8.26) and the assumption that N(ads) is the predominant surface species, it permits θ_N to be expressed in terms of ambient pressures of ammonia and hydrogen.

In a notable theoretical treatment of this reaction in 1940, Temkin used the isotherm that has come to be called by his name. At that time it was a purely empirical relationship, but since then it has been demonstrated that this isotherm represents the behaviour expected if there is considerable intrinsic heterogeneity of surface sites, with a linear distribution of the heat of adsorption from a maximum to a minimum value. There are also related variations in the activation energies of adsorption and desorption over the surface sites, with the former increasing and the latter decreasing as the heat of adsorption decreases, as illustrated in Figure 8.4.

Given this behaviour, the equations for the rates of adsorption and desorption are as follows,

$$R_{ads} = k_a p\, e^{-g\theta} \tag{8.28}$$

$$R_{des} = k_d\, e^{h\theta} \tag{8.29}$$

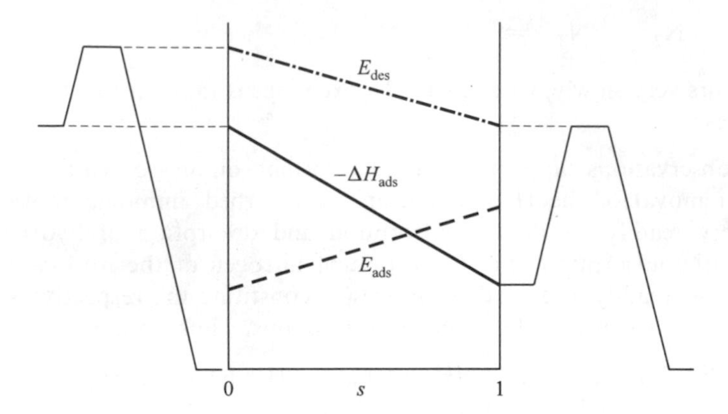

Fig. 8.4 Illustration of the linear variation, over the adsorption sites on the catalyst surface, of the heat of adsorption, ΔH_{ads}, and the activation energies for adsorption (E_{ads}) and desorption (E_{des}), that leads to adsorption behaviour as described by the Temkin equation.

where p denotes the pressure of the adsorbate, θ is the fraction of the surface sites it occupies and k_a and k_d are constants. The constants g and h are proportional to the rates of variation of E_a and E_d over the surface sites. Equating (8.28) and (8.29) we obtain the isotherm, which may be written,

$$\theta = \frac{1}{f} \ln ap \tag{8.30}$$

where $f = (g + h)$ and all these quantities are positive, and $a = k_a/k_d$.

Thus the rate of ammonia decomposition will be given by:

$$\begin{aligned} \text{Rate} = R_{des} &= k_d e^{h\theta} \\ &= k_d (a p^*_{N_2})^{h/f} \\ &= k_d \left(\frac{a}{K}\right)^{h/f} \left(\frac{p^2_{NH_3}}{p^3_{H_2}}\right)^{h/f} \end{aligned} \tag{8.31}$$

So the theory predicts that the indices of NH_3 and H_2 in the rate equation should be in the ratio of 2:−3. The actual values would depend on the behaviour of the catalyst, but the ratio h/f should lie between zero and one. This shows very good agreement with experiment, with h/f showing some variation between different metals, perhaps depending on the form of the metal (powder, wire, foil, film) and the promoters used.

On palladium and platinum catalysts the hydrogen pressure exponent is often found to be more negative than equation (8.31) would require. This probably reflects the fact that on these metals, nitrogen is not nearly so strongly adsorbed as on transition metals of earlier Groups, so that adsorption of nitrogen may well be inhibited by that of hydrogen.

8.8 The criterion for a good catalyst

The question of how one can hope to identify a good catalyst for a particular reaction may be addressed in the context of the oxidation of carbon monoxide. There are two bonds in the molecules involved in this reaction, of which one, the O=O bond, needs to be broken for reaction to occur. So the simplest way in which the reaction may be facilitated is that this bond rupture is achieved when O_2 is chemisorbed on the catalyst.

A Group VI transition metal, like molybdenum, could readily accomplish this, but the bond so formed, between an O atom and a surface metal atom, would be so strong that this species would be of no use in oxidising CO to CO_2. The merit of palladium and platinum is that they interact strongly enough with O_2 to cause its dissociation but not so strongly that O(ads) has difficulty in reacting with CO.

Figure 8.3 is relevant here, even though it was constructed in relation to a slightly different reaction system. The more strongly adsorbed the intermediate, the greater the barrier that has to be surmounted for the reaction to take place.

In regard to ammonia synthesis, the story is very similar. The catalyst must be capable of dissociatively chemisorbing nitrogen, which can then readily be hydrogenated. Some transition metals from the earlier groups, such as vanadium, molybdenum and tungsten, interact very strongly with N_2, but because they bond the N atoms so strongly they are not good catalysts. In later Groups, metals like platinum and palladium have difficulty in dissociatively chemisorbing N_2, but any adsorbed N atoms would be readily hydrogenated. The optimum catalyst for ammonia synthesis might be expected to lie between these groups of examples, which does indeed seem to be the case, with iron and ruthenium among the best metals for this purpose.

Another catalytic reaction which it is instructive to consider in this connection is the dehydrogenation of formic acid:

$$HCO_2H \quad = \quad H_2 + CO_2 \qquad (8.32)$$

On some metals, such as gold and silver, the reaction displays first order kinetics, but on others such as nickel it is essentially zero order. By comparison with the hypothetical reaction considered in Section 8.4, it would then look as if the surface of a nickel catalyst is almost completely covered ($\theta \approx 1$), which implies strong adsorption, whereas on gold and silver it would seem that the surface sites are mostly vacant, implying weak adsorption.

The understanding of the mechanism of this reaction was assisted by complementary studies which showed that when formic acid adsorbed on silica-supported metal particles, a vibration frequency corresponding to the formate ion is seen. This must mean that on adsorption, formic acid sheds its acidic proton and becomes an adsorbed formate ion. Its subsequent behaviour is then expected to resemble that of a metal formate.

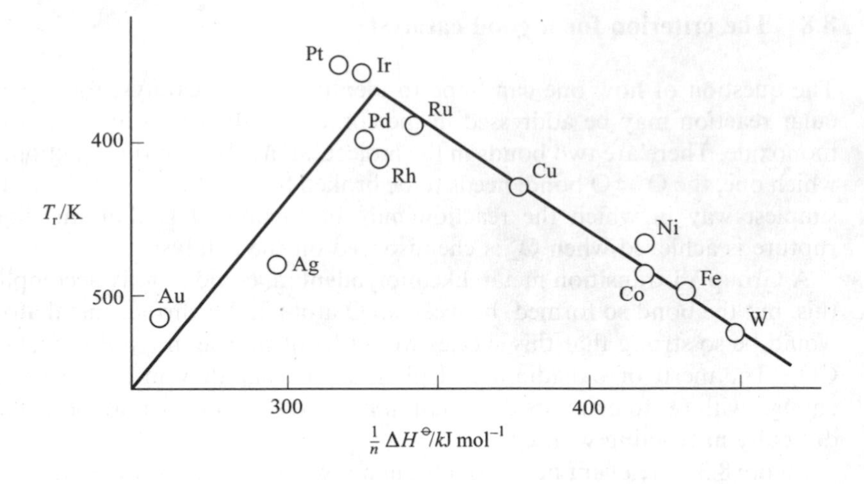

Fig. 8.5 Plot of T_r (the rate at which a specified rate of reaction is achieved per unit surface area of the metal) against the standard enthalpy of formation, per formate ion, of the metal formate. (The data are taken from Table 4 in Fahrenfort, J., L. L. van Reijen and W. M. H. Sachtler, *Z. Elektrochemie,* 1960, **64,** 216– 24, and from Table 1 in Sachtler, W. M. H. and J. Fahrenfort, *Actes du 2eme Congrès International de Catalyse*, Paris, Editions Technip, pp. 831–52.)

One of the most interesting correlations of catalytic activity is shown in Figure 8.5. The abscissa is the standard enthalpy of formation of the metal formate, per formate ion involved. The ordinate serving as the efficiency of the catalyst for reaction (8.32), is the temperature at which a chosen rate of reaction per unit surface area is achieved, and an inverted scale is used since a low temperature signifies a good catalyst. The volcano-shaped plot indicates that the optimum catalyst lies between, on the one hand, gold and silver and, on the other, nickel. The metals at the extreme right have so large a heat of formation of their formate that one must expect much stronger adsorption to the metal surface than is ideal. The optimum conditions are obviously achieved around platinum and iridium, where presumably the adsorption is not so strong as with nickel, but it is stronger than with gold and silver so that there is a greater usage of the available surface sites.

8.9 Catalysis in industry

The foregoing account of kinetic aspects of heterogeneous catalysis refers, for the most part, to laboratory studies of the various reactions involved. The commercial significance has been referred to, and catalytic processes are widely employed in the manufacture both of bulk chemicals and of fine chemicals. In the first group there are various cracking processes in the petrochemical industry, numerous hydrogenation and dehydrogenation processes, oxidation reactions of both simple inorganic and simple organic compounds, in addition to the ammonia synthesis process already mentioned.

In using catalysts industrially, is not possible to work in ultra-clean conditions and to use a freshly prepared surface every day. Catalysts are required to have a substantial working life and to retain their effectiveness. One danger is that, under the conditions necessary for the reaction, the catalyst will undergo sintering so that its surface area is progressively reduced with a concomitant fall in its activity.

In some cases, in order to combat the tendency to sintering, substances called promoters are added to the catalyst. For example, Brunauer and Emmett, in their work at the US Nitrogen Fixation Laboratory, reported on the adsorption and the catalytic behaviour of various commercial promoted iron catalysts. In the singly promoted catalysts, a few percent Al_2O_3 was added to and mixed with Fe_2O_3, and this mixture was heated in a stream of hydrogen gas to generate the catalyst. The doubly promoted catalysts had both Al_2O_3 and K_2O added.

Another hazard for an industrial catalyst is poisoning, which means that an impurity attaches itself firmly to the adsorption sites with the result that these are no longer available for the catalytic reaction. It is desirable to remove such impurities from the feedstock or to use a source in which no such impurity is present. In some cases, it is possible to regenerate a poisoned catalyst, so that if the reaction process is suspended, the catalyst may be treated and restored *in situ*. The beneficial effects of catalyst promoters tend not to be fully reported in the scientific literature, but it is understood that these are not necessarily restricted to combating sintering. They may also help to thwart the effects of catalyst poisons or to increase the intrinsic activity of the original catalyst. For example, the presence of a very small amount of potassium on the surface of iron crystals causes an increase in the sticking coefficient for N_2. That is, the added potassium causes N_2 to be dissociatively chemisorbed more readily, thus accelerating the slow step in ammonia synthesis.

A more recent development in regard to industrial catalysis has been the introduction of bifunctional (or multifunctional) catalysts, capable of expediting reactions of more than one type. While this might perhaps be achieved by using two separate catalysts, having the two different types of site present together on the surface of one catalyst means that mere surface mobility, and not desorption and re-adsorption steps, suffices to take a molecule from the site of a reaction step of one type to a site for the other type. If the process involved requires several reaction steps of one type interspersed with those of another, then the benefit of a bifunctional catalyst may be quite substantial.

An example of the use of a bifunctional catalyst occurs in oil refining, where the feedstock has the desired number of carbon atoms but is a linear alkane, and it is desired to introduce branched chains, 5- and 6-membered

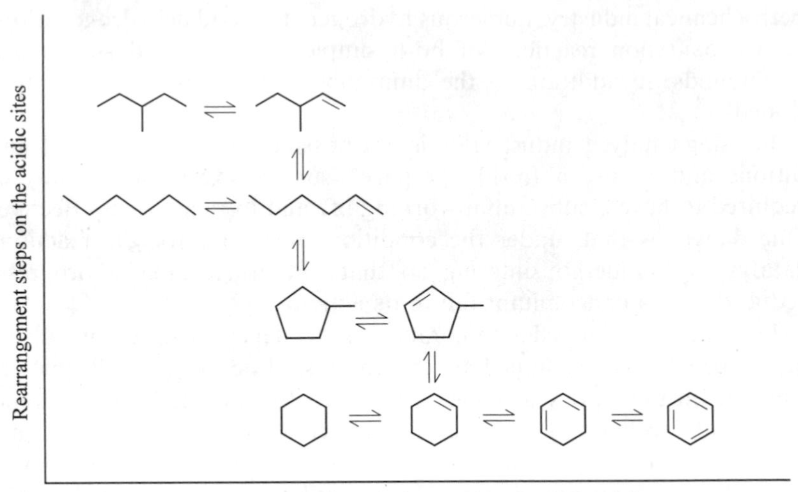

Fig. 8.6 Illustration of the reaction processes on a bifunctional catalyst, where the rearrangement steps on the surface of the acidic oxide are shown on the vertical axis and the hydrogenation and dehydrogenation steps on the platinum surface are shown on the horizontal axis.

rings, and both olefinic and aromatic unsaturation. To achieve these, it is necessary to have a catalyst for rearrangement reactions and a catalyst for dehydrogenation and hydrogenation. A special bifunctional catalyst, in which less than 1% of platinum is very finely dispersed on the surface of an acidic oxide such as alumina, can achieve both. An illustration of the reactions involved, showing how rearrangement processes on the acid sites are interspersed with dehydrogenation steps on the platinum sites, is shown in Figure 8.6.

Suggested reading

Gasser, R. P. H., 1985. *An Introduction to Chemisorption and Catalysis by Metals.* Oxford: Clarendon Press.

White, M. G., 1990. *Heterogeneous Catalysis.* Englewood Cliffs, NJ: Prentice-Hall.

Gates, B. C., 1992. *Catalytic Chemistry.* New York: Wiley.

Bond, G. C., 1987. *Heterogeneous Catalysis: Principles and Applications*, 2nd edition. London: Oxford University Press.

Thomas, J. M. and W. J. Thomas, 1967. *An Introduction to the Principles of Heterogenous Catalysis.* London: Academic Press.

Srivastava, R. D., 1988. *Heterogeneous Catalytic Science.* Boca Raton, FL: CRC Press.

Kemball, C., 1960. Progress in the study of heterogeneous catalysis. *Proceedings of the Chemical Society*, 264–74.

Engel, T. and G. Ertl, 1979. Elementary steps in the catalytic oxidation of carbon monoxide on platinum metals. *Advances in Catalysis*, **28**, 1–78.
Ertl, G., 1983. Kinetics of chemical processes on well-defined surfaces. In *Catalysis, Science and Technology*, Vol. 4. (eds Anderson, J. R. and M. Boudart), pp. 209–82. Berlin: Springer-Verlag.
Somorjai, G. A., 1994. *Introduction to Surface Chemistry and Catalysis*. New York: Wiley.
Satterfield, C. N., 1991. *Heterogeneous Catalysis in Industrial Practice*, 2nd edition. New York: McGraw-Hill.

Problems

8.1 When I_2(g) was allowed to come to an equilibrium with its adsorbed phase on a tungsten surface, the number of molecules adsorbed, n, varied with equilibrium pressure p as follows:

$p/10^{-3}$torr	$n/10^{17}$
0.50	0.499
0.99	0.649
1.48	0.759
1.95	0.851
3.12	0.954
4.13	1.121
5.13	1.190
6.10	1.245
6.78	1.277
7.39	1.325
9.25	1.346

Show that these data fit the Langmuir adsorption isotherm appropriate for dissociative chemisorption and evaluate the number of I_2 molecules needed to form a monolayer. (Data from Campbell, J. S., R. L. Moss and C. Kemball, *Trans Faraday Soc.*, 1960, **56**, 1481–6.)

8.2 The effect of added oxygen on the rate of deuterium exchange of n-hexane on palladium is shown below, for experiments at 359 K.

p_{O_2}	0	0.43	1.06	3.78
k_ϕ/min^{-1}	7.39	3.01	2.22	1.35

Show that these data may be interpreted on the basis that the dissociative chemisorption of O_2 follows the appropriate Langmuir isotherm and the exchange reaction occurs on that fraction of the surface not covered by oxygen. (Data from Gault, F. G. and C. Kemball, *Trans Faraday Soc.*, 1961, **57**, 1781–94.)

8.3 The vapour phase reaction, over an alumina catalyst,

$$CH_3CH_2CH_2OH \;=\; CH_3CH{:}CH_2 + H_2O$$

shows the following kinetic behaviour:

$$\frac{d[\text{propene}]}{dt} = k\,\frac{p_{1\text{-propanol}}}{p_{H_2O}}$$

Propose a Langmuir–Hinshelwood mechanism which accounts for the form of this equation.

8.4 Consider the following variation of the Langmuir–Hinshelwood mechanism for catalysed reactions. Reactants A and B both undergo adsorption on single sites on the catalyst surface. These two adsorbed species react together to produce chemisorbed C. The slow step of the reaction is the desorption of C.

Derive an equation for the rate of this reaction in terms of the pressures of A and B, assuming that the reaction is far from equilibrium. Taking the available options regarding whether species are strongly or weakly adsorbed, list the various extreme possibilities for the reaction order with respect to A and B.

9 Homogeneous catalysis

In a homogeneous phase, if a substance not involved in the balanced equation for the reaction is added, this may make possible a succession of steps leading to the same reaction products and the regeneration of the added substance. The rate of the reaction has thus been enhanced by a catalytic mechanism which has many of the characteristics of a chain reaction.

For some reactions, a reaction product functions as an active catalyst, so that the rate increases over the initial stages as the concentration of this product/catalyst builds up. This causes the time profile of reactant concentration to have the distinctive sigmoid shape which is characteristic of autocatalysis. We shall show that a mechanism in which there are two successive autocatalytic steps may demonstrate periodic oscillations in the concentration of the reactive intermediates involved. Such reactions, in which these intermediates appear to flout the Principle of Stationary States, are widely, if misleadingly, called "oscillating reactions".

It may also be stressed that, by definition, a catalyst is a *substance*. Heat and light may assist in achieving a reaction, but they cannot be catalysts.

9.1 Gas phase catalysis

An illustrative example of gas phase catalysis is provided by the effect of molecular bromine on the decomposition of ozone, described by the stoichiometric equation:

$$2O_3 = 3O_2 \tag{9.1}$$

The mechanism involved is believed to be as follows:

$$\left.\begin{aligned} Br_2 &\xrightarrow{1} Br^{\cdot} + Br^{\cdot} \\ Br^{\cdot} + O_3 &\xrightarrow{2} BrO^{\cdot} + O_2 \\ BrO^{\cdot} + O_3 &\xrightarrow{3} Br^{\cdot} + O_2 + O_2 \\ Br^{\cdot} + O_3 &\xrightarrow{4} BrO_3 \end{aligned}\right\} \tag{9.2}$$

This is a chain mechanism, with $Br^{\cdot}$ and $BrO^{\cdot}$ as the chain carriers, step 1 as initiation and step 4 as termination. Application of the Principle of Stationary States leads to the following equations:

$$\frac{d[Br^{\cdot}]}{dt} = 2k_1[Br_2] - k_2[Br^{\cdot}][O_3] + k_3[BrO^{\cdot}][O_3] - k_4[Br^{\cdot}][O_3] = 0 \quad (9.3)$$

$$\frac{d[BrO^{\cdot}]}{dt} = k_2[Br^{\cdot}][O_3] - k_3[BrO^{\cdot}][O_3] = 0 \quad (9.4)$$

Adding these equations and rearranging, we have:

$$[Br^{\cdot}] = \frac{2k_1[Br_2]}{k_4[O_3]} \quad (9.5)$$

Using equation (9.4), this gives the steady state concentration of the other chain carrier:

$$[BrO^{\cdot}] = \frac{2k_1k_2[Br_2]}{k_3k_4[O_3]} \quad (9.6)$$

So, for the rate of reaction (9.1) we have:

$$\begin{aligned} \frac{1}{3}\frac{d[O_2]}{dt} &= \frac{1}{3}\{k_2[Br^{\cdot}][O_3] + 2k_3[BrO^{\cdot}][O_3]\} \\ &= \frac{2k_1k_2}{k_4}[Br_2] \end{aligned} \quad (9.7)$$

Thus the rate of the bromine-assisted reaction is proportional to the concentration of Br_2. There will also be some reaction not involving bromine, so that the qualitative dependence of the rate on Br_2 concentration will be as sketched in Figure 9.1. This is a general characteristic of a catalysed reaction.

However, strictly speaking, mechanism (9.2) cannot quite meet the criteria for catalysis, in that the bromine atoms end up as BrO_3 and not as Br_2. But a distinction can easily be made between the role of Br_2 in this experiment and that of an initiator in a free radical addition polymerisation reaction (see Section 7.3). In the latter, where the initiator I yields two free radical species $R^{\cdot}$, the only role for this radical $R^{\cdot}$ is to react with a monomer molecule, so that $R^{\cdot}$ has been removed before the start of the propagation steps. In mechanism (9.2), $Br^{\cdot}$ is one of the two free radical chain carriers and fails to re-form Br_2 only because one of the propagation steps has an alternative outcome which represents termination.

These examples illustrate the role of a catalyst in providing an alternative mechanism for the reaction which serves as a speedier route to the reaction products. Two points may be noted:

(i) In general, the reason the catalysed reaction is faster is that the alternative path has a lower activation energy than that of the uncatalysed reaction. Evaluation of the Arrhenius activation energies for the two will usually confirm this.

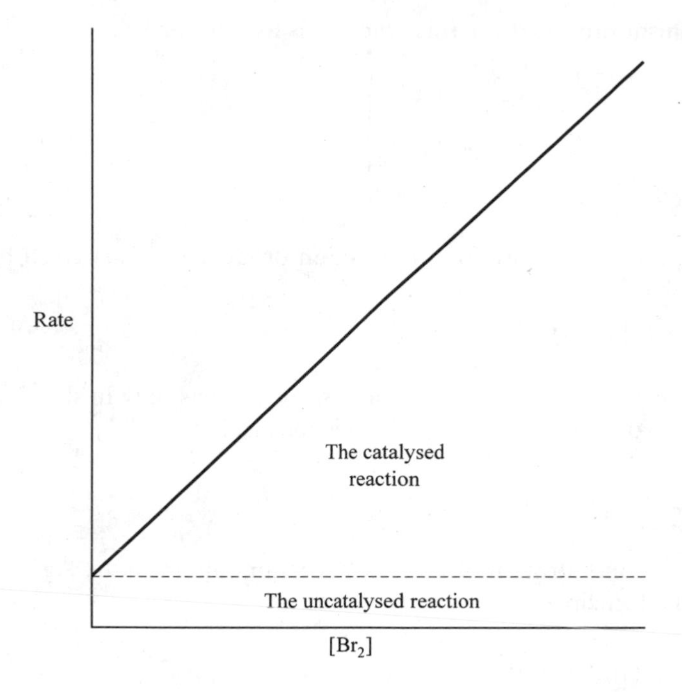

Fig. 9.1 Diagram showing how the rate of reaction (9.2) will vary with the concentration of Br_2, with contributions from the catalysed and an uncatalysed reaction mechanism.

(ii) By contrast an added substance which slows down a reaction must operate in a totally different way, by interfering with the mechanism of the usual reaction. For that reason, it is best called a reaction inhibitor.

9.2 Catalysis in solution

An example of homogeneous catalysis in solution is provided by the redox reaction between ceric and thallous ions:

$$2Ce^{4+} + Tl^{+} = 2Ce^{3+} + Tl^{3+} \tag{9.8}$$

This reaction occurs extremely slowly, but it is catalysed by Ag^{+} at a rate proportional to its nominal concentration.

The mechanism proposed for this reaction is as follows:

$$\left.\begin{aligned} Ag^{+} + Ce^{4+} &\underset{-1}{\overset{1}{\rightleftarrows}} Ag^{2+} + Ce^{3+} \\ Ag^{2+} + Tl^{+} &\overset{2}{\rightarrow} Ag^{+} + Tl^{2+} \\ Tl^{2+} + Ce^{4+} &\overset{3}{\rightarrow} Tl^{3+} + Ce^{3+} \end{aligned}\right\} \tag{9.9}$$

On this basis, the rate of forming the reaction product, Tl^{3+}, is given by:

$$\frac{d[Tl^{3+}]}{dt} = k_3[Tl^{2+}][Ce^{4+}] \tag{9.10}$$

This ion Tl^{2+} is formed only in step 2, and since it reacts only in step 3, the rates of these two steps must be equal. This means:

$$\frac{d[Tl^{3+}]}{dt} = k_2[Ag^{2+}][Tl^{+}] \tag{9.10a}$$

To evaluate the concentration of the other reaction intermediate, Ag^{2+}, we need to use the Principle of Stationary States:

$$\frac{d[Ag^{2+}]}{dt} = k_1[Ag^{+}][Ce^{4+}] - k_{-1}[Ag^{2+}][Ce^{3+}] - k_2[Ag^{2+}][Tl^{+}] = 0 \tag{9.11}$$

$$\therefore [Ag^{2+}] = \frac{k_1[Ag^{+}][Ce^{4+}]}{k_{-1}[Ce^{3+}] + k_2[Tl^{+}]} \tag{9.12}$$

Thus we have:

$$\begin{aligned} \frac{d[Tl^{3+}]}{dt} &= \frac{k_1k_2[Ag^{+}][Ce^{4+}][Tl^{+}]}{k_{-1}[Ce^{3+}] + k_2[Tl^{+}]} \\ &= \frac{k_1[Ag^{+}][Ce^{4+}]}{(k_{-1}/k_2)[Ce^{3+}]/[Tl^{+}] + 1} \end{aligned} \tag{9.10b}$$

The value of k_{-1}/k_2 is such that neither term in the denominator predominates sufficiently to justify ignoring the other. However, in the late stages of the reaction the first term must predominate. The rate then approximates to:

$$\frac{d[Tl^{3+}]}{dt} \approx \frac{k_1k_2}{k_{-1}}[Ag^{+}][Ce^{4+}][Tl^{+}]/[Ce^{3+}] \tag{9.10c}$$

Silver(I) ion also acts as a catalyst for the reaction:

$$2VO_3{}^{+} + S_2O_8{}^{2-} = 2VO^{2+} + 2SO_4{}^{2-} + 2O_2 \tag{9.13}$$

The rate of the catalysed reaction is independent of the concentration of peroxovanadium ion, $VO_3{}^{+}$. The mechanism is believed to be:

$$\left.\begin{array}{rcl} Ag^+ + S_2O_8{}^{2-} & \xrightarrow{1} & Ag^{2+} + SO_4{}^- + SO_4{}^{2-} \\ Ag^+ + SO_4{}^- & \xrightarrow{2} & Ag^{2+} + SO_4{}^{2-} \\ Ag^{2+} + VO_3{}^+ & \xrightarrow{3} & Ag^+ + VO_3{}^{2+} \\ VO_3{}^{2+} & \xrightarrow{4} & VO^{2+} + O_2 \end{array}\right\} \quad (9.14)$$

Step 1 is rate-determining here, producing one Ag^{2+} ion directly and a further Ag^{2+} ion by the subsequent reaction of the $SO_4{}^-$ ion also produced in step 1. Thus the rate of the catalysed reaction is given by:

$$\frac{d[VO^{2+}]}{dt} = 2k_1[Ag^+][S_2O_8^{2-}] \quad (9.15)$$

Since step 1 does not involve the $VO_3{}^+$ ion, the reaction shows zero order kinetics with respect to it. As usual, the catalysed reaction shows first order dependence on the catalyst, Ag^+ ion.

A rather different example of homogeneous catalysis in solution is provided by the aqueous hydrolysis of methyl bromide,

$$CH_3Br + H_2O = CH_3OH + HBr \quad (9.16)$$

which is catalysed by iodide ion. The mechanism is as follows. First, iodide ion reacts with CH_3Br in an S_N2 process, in which I^- acts as a nucleophile, to give CH_3I:

$$CH_3Br + I^- \xrightarrow{1} CH_3I + Br^- \quad (9.17)$$

The second step involves another S_N2 process in which a water molecule reacts with methyl iodide:

$$CH_3I + H_2O \xrightarrow{2} CH_3OH + I^- + H^+ \quad (9.18)$$

The presence of iodide ion causes an acceleration only because steps 1 and 2 are both faster than the direct S_N2 attack of a water molecule on CH_3Br. However, with a 100-fold excess of iodide ion, the catalysed reaction is faster by a factor of only about 5.

One characteristic of the catalysis of this reaction by adding iodide is that, in providing a new mechanism, the Gibbs energy of activation has been lowered, but there are consequently two barriers to be overcome in place of the original one. However, the maximum barrier height is lower, and it is for that reason that iodide is an effective catalyst in this reaction.

9.3 Catalysis by acids

The catalytic role of acids has been known since the time of Berzelius. The topic was studied in some detail by Ostwald in the 1880s.

One example is the halogenation of acetone (2-propanone),

$$CH_3COCH_3 + X_2 = CH_3COCH_2X + HX \quad (9.19)$$

where, in acidic solution, the rate of the reaction is proportional to the concentrations both of acetone and of H^+, but is independent of the concentration of the halogen, X_2. Thus it may be deduced that X_2 does not participate in the reaction sequence until after the rate-determining step.

Like any other ketone bearing an α-hydrogen atom, acetone has a keto and an enol form, with the former predominant. The enol form readily reacts with X_2, whereas the keto form does not. The first stage in reaction (9.19) is the conversion of keto to enol, a process which is catalysed by acids and by bases.

In acid solution, the first stage in enolisation is the rapid, reversible addition of a proton to the carbonyl oxygen, to give a species which, for short, will be denoted as **YH**$^+$:

$$CH_3COCH_3 + H_3O^+ \overset{1}{\rightleftharpoons} \underset{(\mathbf{YH^+})}{CH_3\overset{\overset{+}{O}H}{\overset{||}{C}}CH_3} + H_2O \qquad (9.20)$$

Thus we have:

$$K_1 = \frac{[\mathbf{YH}^+]}{[\text{keto}][H_3O^+]} \qquad (9.21)$$

To form the enol from **YH**$^+$, a proton must be removed from one of the methyl groups. We may conceive of different acceptors of this proton:

$$\left.\begin{aligned} &\mathbf{YH}^+ + OH^- \xrightarrow{2} CH_3\overset{OH}{\overset{|}{C}}{:}CH_2 + H_2O \\ &\mathbf{YH}^+ + H_2O \xrightarrow{3} CH_3\overset{OH}{\overset{|}{C}}{:}CH_2 + H_3O^+ \\ &\mathbf{YH}^+ + A^- \xrightarrow{4} CH_3\overset{OH}{\overset{|}{C}}{:}CH_2 + HA \end{aligned}\right\} \qquad (9.22)$$

The rate of forming the enol is the sum of the rates of these three steps.

$$\frac{d[\text{enol}]}{dt} = k_2[\mathbf{YH}^+][OH^-] + k_3[\mathbf{YH}^+] + k_4[\mathbf{YH}^+][A^-] \qquad (9.23)$$

Substituting from equation (9.20) for the concentration of the protonated form of acetone, **YH**$^+$, we have:

$$\frac{\mathrm{d[enol]}}{\mathrm{d}t} = [\mathrm{keto}][\mathrm{H_3O^+}]K_1\{k_2[\mathrm{OH^-}] + k_3 + k_4[\mathrm{A^-}]\} \tag{9.23a}$$

The product of $[H_3O^+]$ and $[OH^-]$ is of course equal to K_w, the ionic product of water. Also, for the weak acid, HA, whose dissociation constant is K_a, the product of $[H_3O^+]$ and $[A^-]$ is equal to $K_a[HA]$. Thus the rate equation may be written:

$$\frac{\mathrm{d[enol]}}{\mathrm{d}t} = [\mathrm{keto}]K_1\{k_2K_w + k_3[\mathrm{H_3O^+}] + k_4K_a[\mathrm{HA}]\} \tag{9.23b}$$

The last step in the halogenation reaction (9.19) is the reaction of X_2 with the enol form:

$$\mathrm{CH_3C(OH){:}CH_2 + X_2 \xrightarrow{5} CH_3COCH_2X + HX} \tag{9.24}$$

For Br_2 and I_2, this is rapid, with the rate of step 5 controlled by the availability of enol. Thus the rate of halogenation is given by equation (9.23) and is independent of the concentration and the identity of the halogen. The concentration of the keto form is essentially the same as that of acetone, so we may write the rate equation, at its simplest, as:

$$\frac{\mathrm{d[CH_3COCH_2X]}}{\mathrm{d}t} = [\mathrm{CH_3COCH_3}]\{k_o + k_{\mathrm{H^+}}[\mathrm{H_3O^+}] + k_{\mathrm{HA}}[\mathrm{HA}]\} \tag{9.23c}$$

here $k_o = k_2K_1K_w$, $k_{H^+} = k_3K_1$ and $k_{HA} = k_4K_1K_a$.

In equation (9.23c), the constant k_o is often described as arising from the uncatalysed reaction. More precisely, it represents the contribution of the solvent to the catalysed reaction, in that steps 1 and 2 can occur even at pH 7. The yield of enol from step 2 is independent of pH, as shown in equation (9.23b).

The term $k_{H^+}[H_3O^+]$ represents the contribution to the catalysed reaction from step 3 of mechanism (9.22), and is called specific acid catalysis. The term $k_{HA}[HA]$, arising from step 4, represents the contribution arising from the participation of the conjugate base, A^-. This is termed general acid catalysis, and where this is evident it means that A^- is far more proficient than the more abundant molecules of solvent in accepting a proton from a methyl group of the ion, $\mathbf{Y}H^+$.

Experimental determination of the rate constants k_o, k_{H^+} and k_{HA} is complicated by the fact that one of the reaction products, HX, is a strong acid. In the absence of any weak acid, the experimental pseudo-first order rate constant k' is equal to $(k_o + k_{H^+}[H^+])$ and the individual values may be found as the intercept and slope of a plot of k' against the H^+ concentration.

If a weak acid HA is thought to be a catalyst, then experiments should be carried out using HA/A^- buffers of varying concentration but constant pH. The plot of the experimental rate constant k' against the HA concentration is then found either to be flat, indicating that there is no such catalysis, or to have a positive slope equal to k_{HA}.

General acid catalysis is not invariably found for all acid-catalysed reactions. In some cases the reason appears to be that the rate-determining step is a unimolecular dissociation of the protonated species, so that there is no proton-accepting role for a species such as A^-. This occurs in the hydrolysis of an acetal,

$$\mathrm{RHC(OR')_2} + \mathrm{H_2O} = \mathrm{RHC{=}O} + 2\,\mathrm{R'OH} \qquad (9.25)$$

whose mechanism is believed to be as follows:

$$\left.\begin{array}{l}
\mathrm{RHC(OR')_2} + \mathrm{H_3O^+} \underset{-1}{\overset{1}{\rightleftharpoons}} \mathrm{RHC(OR')(O^+HR')} + \mathrm{H_2O} \\
\mathrm{RHC(OR')(O^+HR')} \xrightarrow{2} \mathrm{RHC{=}O^+R'} + \mathrm{R'OH} \\
\mathrm{RHC{=}O^+R'} + 2\mathrm{H_2O} \xrightarrow{3} \mathrm{RHC{=}O} + \mathrm{R'OH} + \mathrm{H_3O^+}
\end{array}\right\} \qquad (9.26)$$

The rate of step 2, which is rate-determining, reflects only the concentrations of acetal and of H_3O^+, regardless of whether the solution contains strong or weak acids.

9.4 Catalysis by bases

The halogenation of acetone is also catalysed by bases. The mechanism is necessarily different from acid catalysis and the first and rate-determining step is the removal of a proton by a base. Other bases in addition to OH^- may participate:

$$\left.\begin{array}{l} CH_3COCH_3 + OH^- \xrightarrow{6} CH_3C^-OCH_2 + H_2O \\ CH_3COCH_3 + H_2O \xrightarrow{7} CH_3C^-OCH_2 + H_3O^+ \\ CH_3COCH_3 + B^- \xrightarrow{8} CH_3C^-OCH_2 + BH \end{array}\right\} \quad (9.27)$$

The halogen now reacts rapidly with this anion to give haloacetone:

$$CH_3C^-OCH_2 + X_2 \xrightarrow{9} CH_3COCH_2X + X^- \quad (9.28)$$

This leads to the rate expression,

$$\text{Rate} = [CH_3COCH_3]\,\{k_7 + k_6\,[OH^-] + k_8\,[B^-]\} \quad (9.29)$$

and the individual rate constants k_7, k_6 and k_8 may be evaluated by procedures similar to those described above for the identification of the rate constants in the acid-catalysed reaction.

Where a reaction undergoes specific acid and specific base catalysis, then the relevant parts of equations (9.23c) and (9.29) are both applicable and these may be combined in one expression for the apparent first order rate constant, k', of the catalysed reaction:

$$\begin{aligned} k' &= \{k_o + k_{H^+}[H^+] + k_{OH^-}[OH^-]\} \\ &= \{k_o + k_{H^+}[H^+] + k_{OH^-}K_w[H^+]^{-1}\} \end{aligned} \quad (9.30)$$

A plot of $\log_{10}k'$ against pH may then have three linear portions:

(i) at low pH, the second term will predominate, so the slope will be –1.
(ii) at high pH, the third term will predominate, so the slope will be + 1.
(iii) between these two ranges the first term will be the largest, so the slope will be zero.

This is illustrated in Figure 9.2.

In the examples of catalysis by bases shown in mechanism (9.27), the only role for the base has been to accept a proton from the substrate. However, it is possible that a base may become more closely involved in the mechanism of a reaction, so that at the rate-determining step it is attached to the substrate by a covalent bond. The kinetic consequence is that the pseudo-first order rate constant varies with the base concentration just as in equation (9.29). However, the value of the relevant rate constant may be much greater than that expected if the base were merely acting as a proton acceptor.

A notable example of this type of catalysis by a base, known as nucleophile catalysis, was discovered in regard to the hydrolysis of *p*-nitrophenyl acetate, in which imidazole was found to be *ca* 10^3 times more effective than HPO_4^{2-}, which has the same basicity constant. The mechanism is believed to be as follows:

(where **Im** ≡ —N⟨imidazole ring⟩=N)

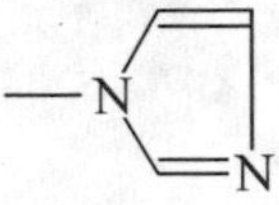

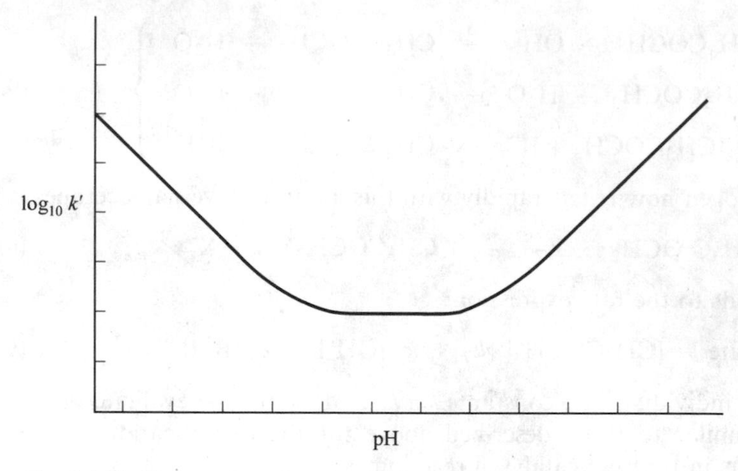

Fig. 9.2 A diagram based on equation (9.30) showing how $\log_{10}k'$ varies with pH, with linear portions corresponding to the predominance of the acid-catalysed, solvent-catalysed and base-catalysed mechanisms.

$$\left.\begin{array}{l}
\mathrm{CH_3{-}C({=}O){-}O{-}Ph} \quad \textbf{ImH}\ \mathrm{H_2O} \overset{1}{\rightleftharpoons} \mathrm{CH_3{-}C(O^-)(\textbf{Im}){-}O{-}Ph}\ \textbf{(Y)} \quad \mathrm{H_3O^+} \\
\mathrm{CH_3{-}C(O^-)(\textbf{Im}){-}O{-}Ph}\ \textbf{(Y)} \overset{2}{\rightarrow} \mathrm{CH_3{-}C({=}O)\textbf{Im}} + \bar{\mathrm{O}}\mathrm{Ph} \\
\mathrm{CH_3{-}C({=}O)\textbf{Im}} + \mathrm{H_2O} \overset{3}{\rightarrow} \mathrm{CH_3CO_2H} + \textbf{ImH}
\end{array}\right\} \qquad (9.31)$$

The first step is believed to be in equilibrium, and the second is then rate-determining. This gives, for the rate of reaction,

$$\begin{aligned}\text{Rate} &= k_2[\mathrm{Y}] \\ &= k_2K_1\,[\text{ester}]\,[\mathrm{ImH}]\end{aligned} \qquad (9.32)$$

so that kinetically the role of imidazole in this reaction is similar to that of OH^- ion in mechanism (9.27) and equation (9.29).

An important point about the role of imidazole in mechanism (9.31) is that, by reacting in this manner with the ester, it facilitates the loss of the phenolate anion in step 2: however, it is vital that *N*-acetylimidazole is then readily hydrolysed, thus completing the formation of the products of solvolysis of the original ester, and regenerating the catalyst. In support of mechanism (9.31), it may be added that during the reaction there is clear spectrophotometric evidence for the existence of *N*-acetylimidazole in the solution.

9.5 Autocatalysis

In the infancy of chemical kinetics, when the first experiments were being done to discover the "laws" of chemical kinetics, one of the reactions studied by Harcourt at Oxford was that between oxalic (ethanedioic) acid and permanganate ion in acid solution. This reaction is a familiar one in freshman undergraduate laboratories, where it is used in a self-indicating titration, carried out after the oxalic acid solution has been heated to *ca* 70°C. Also familiar will be the impression, on making the first addition of permanganate from the burette, that the end-point has been overshot. But in a minute or so, the pink colour has been dispelled and subsequent additions of permanganate react very quickly. In short, the reaction is catalysed by one of its products, Mn^{2+} ion, in whose absence the initial reaction occurs so slowly. Given the lack of development of the topic in 1866, it was an unfortunate choice of reaction for study.

The term used to describe this kinetic phenomenon is autocatalysis. For the reaction,

$$\mathrm{A} = \mathrm{B} + \mathrm{C} \tag{9.33}$$

the simplest possible autocatalytic reaction scheme is:

$$\mathrm{A} + \mathrm{B} \xrightarrow{1} \mathrm{B} + \mathrm{B} + \mathrm{C} \tag{9.34}$$

Denoting the initial concentration of A by a and that after time t by $(a - x)$, the corresponding rate expression is:

$$\frac{\mathrm{d}x}{\mathrm{d}t} = k_1(a - x)x \tag{9.35}$$

It can readily be seen that the rate will be a maximum at $x = a/2$. However, if strictly no product B is present initially, then x is zero and the initial rate is zero: unless a small amount of B is obtained by some means, no reaction can occur.

If a finite amount of B is assumed to be present initially, giving the concentration b, then the rate equation becomes:

$$\frac{dx}{dt} = k_1(a-x)(b+x) \tag{9.35a}$$

Integrating by parts, this gives,

$$\begin{aligned}\int k_1 dt &= \int \frac{dx}{(a-x)(b+x)} \\ &= \frac{1}{a+b}\left\{\int \frac{dx}{(a-x)} + \int \frac{dx}{(b+x)}\right\}\end{aligned} \tag{9.36}$$

which leads to

$$\frac{1}{a+b} \ln\left\{\frac{a(b+x)}{b(a-x)}\right\} = k_1 t \tag{9.37}$$

The implications of equation (9.37) for the concentration of A are shown in Figure 9.3, which shows the sigmoid-shaped curve characteristic of an autocatalytic reaction. The figure also illustrates that the choice of the parameter b is quite influential in regard to the time needed to attain the maximum rate, though it scarcely affects the shape of the curve.

Another device which avoids the difficulty inherent in equation (9.35) (that, strictly speaking, it refers to a reaction which would never start) is to assume two parallel channels of reaction, one simple first order and one autocatalytic:

$$\left.\begin{aligned} A &\xrightarrow{1} B + C \\ A + B &\xrightarrow{2} B + B + C \end{aligned}\right\} \tag{9.38}$$

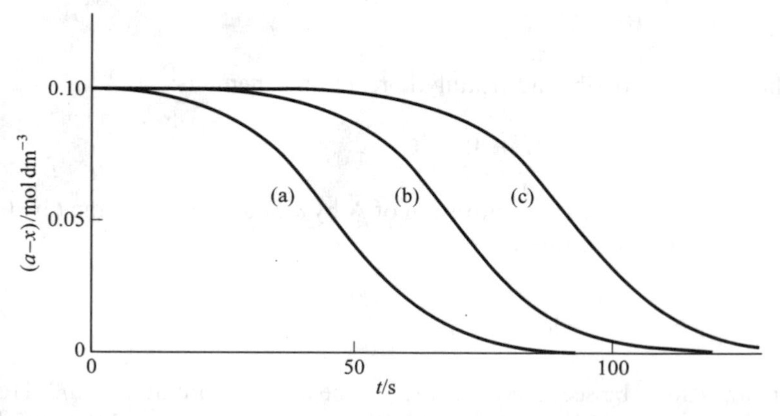

Fig. 9.3 Variation of the concentration of A with time, based on mechanism (9.34) and equation (9.37), for (a) $b = 10^{-3}$ mol dm^{-3}, (b) $b = 10^{-4}$ mol dm^{-3} and (c) $b = 10^{-5}$ mol dm^{-3} taking $a = 0.1$ mol dm^{-3} and $k_1 = 1$ dm^3 mol^{-1} s^{-1}.

Thus the rate equation becomes:

$$\frac{dx}{dt} = k_1(a - x) + k_2x(a - x) \tag{9.39}$$

Separating the variables and using the method of partial fractions, this becomes:

$$\frac{1}{(k_1 + k_2a)} \int \left\{ \frac{1}{a - x} + \frac{k_2}{k_1 + k_2x} \right\} dx = \int dt \tag{9.40}$$

Integrating and using the initial condition that $x = 0$ when $t = 0$, leads to the result,

$$\ln \frac{(k_1 + k_2x)a}{(a - x)k_1} = (k_1 + k_2a)t \tag{9.40a}$$

which clearly reduces to equation (1.15) as k_2 tends to zero. However, the form of this equation is such that, given data from an experiment, the values of k_1 and k_2 are not readily obtained, except by trial and error, or by iteration to discover the ratio of k_1 to k_2.

9.6 Oscillating reactions

The topic of oscillating reactions has been fairly prominent over the past 30 years or so. At the outset, it is important to make clear just what may oscillate.

In a homogeneous reaction,

$$\mathrm{A + B \;=\; C + D} \tag{9.41}$$

the criterion of equilibrium, at constant temperature and pressure, is the position at which the Gibbs energy, G, is minimised, as shown in Figure 9.4. In some cases the approach to equilibrium is extremely slow, because of kinetic limitations. But no matter how rapid the reaction, it can be stated with some certainty that reaction will in no case proceed beyond the point of thermodynamic equilibrium. To do so would infringe the Second Law of Thermodynamics.

So, for reaction (9.41) it is not possible that the concentrations of A and B oscillate around the ξ value corresponding to final equilibrium. But for reaction *intermediates* there is no such constraint. In Chapter 3 we introduced the Principle of Stationary States and in that and subsequent chapters the usefulness of this principle has been apparent. It can now be asserted that although this principle is usually applicable to a reaction intermediate, in special circumstances it has no validity.

It is ironic that the two landmark papers on oscillating reactions were published one just before and one just after the first established use of the Principle of Stationary States. In the first of these, in 1910, Lotka showed that in an autocatalytic reaction the concentration of an intermediate could

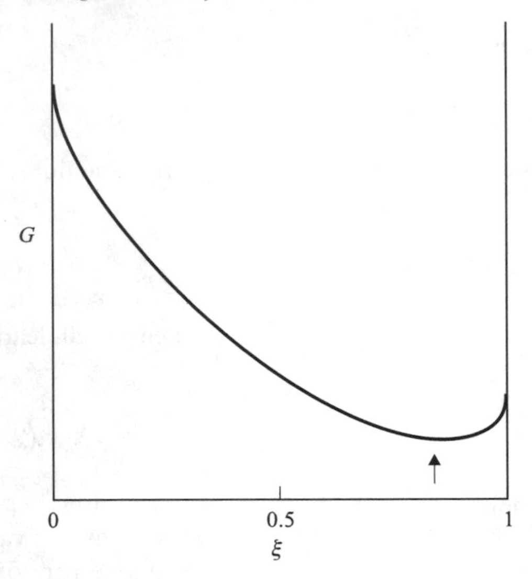

Fig. 9.4 Illustration, for a homogeneous reaction such as (9.41), of how the Gibbs energy of the system, G, varies with ξ, the degree of advancement. The arrow marks the ξ value corresponding to the point of equilibrium.

approach its steady state value through damped oscillations. The second, in 1920, showed that the concentrations of the intermediates may undergo undamped oscillations if there are two successive autocatalytic steps in the reaction. This mechanistic feature is thus a sufficient but not a unique condition for the existence of "an oscillating reaction", and may be used to illustrate what is permissible.

Suppose the overall reaction,

$$A = D \tag{9.42}$$

occurs by the mechanism,

$$\left.\begin{array}{lcl} A + B & \xrightarrow{1} & B + B \\ B + C & \xrightarrow{2} & C + C \\ C & \xrightarrow{3} & D \end{array}\right\} \tag{9.43}$$

where steps 1 and 2 are the autocatalytic steps referred to above. Let us look at the outline of the argument that this system may demonstrate continuing oscillations in the concentrations of the reaction intermediates, B and C.

The equations for the net rate of formation of the intermediates B and C are:

$$\frac{d[B]}{dt} = k_1[A][B] - k_2[B][C] \tag{9.44}$$

$$\frac{d[C]}{dt} = k_2[B][C] - k_3[C] \tag{9.45}$$

If we were invoking the Principle of Stationary States, we would equate each of these to zero, and thus obtain,

$$[C]_{ss} = \frac{k_1}{k_2}[A] = q \tag{9.46}$$

$$[B]_{ss} = \frac{k_3}{k_2} = p \tag{9.47}$$

However, without making the assumption that the Principle is applicable here, we may divide equation (9.44) by equation (9.45). Rearranging, we have:

$$\left\{k_2 - \frac{k_3}{[B]}\right\}d[B] = \left\{\frac{k_1[A]}{[C]} - k_2\right\}d[C] \tag{9.48}$$

By integration, we obtain:

$$k_2[B] - k_3 \ln[B] - k_1[A]\ln[C] + k_2[C] = K \tag{9.49}$$

where K denotes the constant of integration.

The steady state concentrations of B and C, given in equations (9.46) and (9.47), involve only k_1, k_2, k_3 and the concentration of A. These values are thus consistent with equation (9.49), but this is clearly not a unique solution. Let us try the values:

$$[B] = p + x \tag{9.50}$$

$$[C] = q + y \tag{9.51}$$

Substituting these into equation (9.49) we obtain:

$$\begin{aligned} k_2x - k_3\ln(p+x) - k_1[A]\ln(q+y) + k_2y &= K - k_2(p+q) \\ &= K' \end{aligned} \tag{9.52}$$

In order to discover the nature of this relation, it is convenient to accept the limitations $x \ll p$ and $y \ll q$. If we now expand the ln terms, we have,

$$\begin{aligned} \ln(p+x) &= \ln p + \ln(1 + x/p) \\ &\approx \ln p + \frac{x}{p} - \frac{x^2}{2p^2} \end{aligned} \tag{9.53}$$

where higher terms are so close to zero that they may be neglected. And similarly for $\ln(q+y)$. Substituting these into equation (9.52), we obtain:

$$\frac{k_3x^2}{p^2} + \frac{k_1[A]y^2}{q^2} = 2(K' + k_3\ln p + k_1[A]\ln q) \tag{9.54}$$

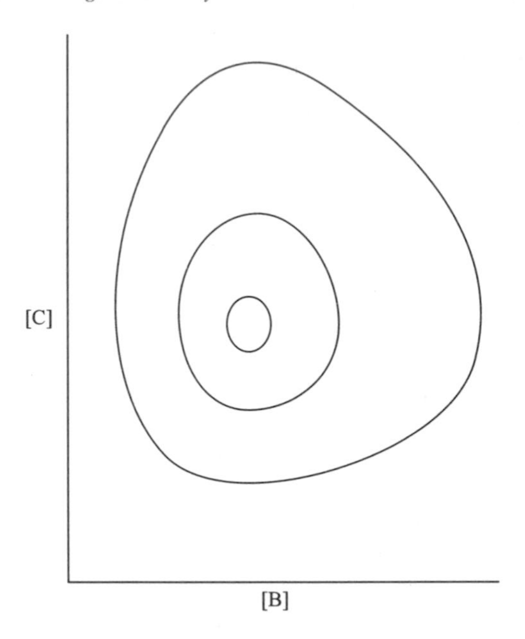

Fig. 9.5 Illustration of the oscillations in the values of $(p + x)$ and $(q + y)$ that are consistent with equation (9.52). For the oscillations of low amplitude the curves closely approximate to ellipses.

The right hand side of this equation is constant. Assuming it is non-zero, (9.54) is the equation for an ellipse. This means that the concentrations of B and C may oscillate around their steady state values, p and q.

Restricting x and y to ranges much less than p and q was for mathematical convenience and is not a requirement of the reaction mechanism. The concentrations of B and C cannot be negative, so the oscillations permitted by equation (9.52) are not entirely elliptical, as is shown in Figure 9.5. Additional analysis shows that the frequency of oscillation is given by:

$$\nu = \{k_1 k_3[\mathrm{A}]\}^{1/2}/2\pi \qquad (9.55)$$

A familiar example of oscillating behaviour is provided by the Belousov–Zhabotinskii reaction, in which the overall reaction involves the oxidation, in aqueous acid solution free of chloride ion, of malonic (1,3-propanedioic) acid to CO_2 and formic (methanoic) acid by bromate ion, with the assistance of catalytic amounts of a metal ion such as Ce^{IV}/Ce^{III}. The oscillations may be detected visually by adding a ferroin indicator, whose colour changes periodically between purple and pale blue, reflecting the repetitive fluctuations in the ratio of $Ce^{IV}:Ce^{III}$. Also, if a bromide-sensitive electrode is dipped into the solution, its electromotive force shows how the Br^- concentration oscillates, as shown in Figure 9.6, at the same frequency as the colour changes of the indicator.

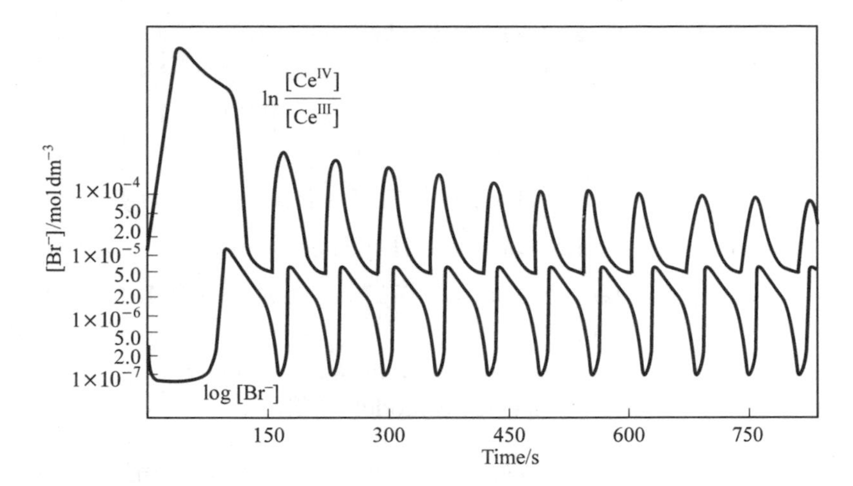

Fig. 9.6 Potentiometric traces of Br^- and of the Ce^{IV}/Ce^{III} ratio as a function of time for a Belousov–Zhabotinskii reaction with initial concentrations: [malonic acid] = 0.13 mol dm^{-3}; $[BrO_3^-]$ = 0.063 mol dm^{-3} ; $[Ce^{IV}]$ = 0.005 mol dm^{-3} and $[H_2SO_4]$ = 0.80 mol dm^{-3}. (Reproduced with permission from Figure 5 of Field, R. J., E. Körös and R. M. Noyes, *J. Am. Chem. Soc.*, 1972, **94**, 8649–64.)

The mechanism of the Belousov–Zhabotinskii reaction involves many steps and several intermediate species, and has been the focus of much attention. The oscillating reaction would seem to be divisible into three phases:

(i) Bromate reacts with malonic acid to produce bromomalonic acid, consuming Br^- in the process:

$$BrO_3^- + 2Br^- + 3CH_2(CO_2H)_2 + 3H^+ = 3BrCH(CO_2H)_2 + 3H_2O \tag{9.56}$$

(ii) The formation of bromomalonic acid by bromate continues, but without involving Br^- ion. Cerous ion is oxidised to ceric:

$$BrO_3^- + 4Ce^{3+} + CH_2(CO_2H)_2 + 5H^+ = BrCH(CO_2H)_2 + 4Ce^{4+} + 3H_2O \tag{9.57}$$

(iii) Finally, ceric ion degradatively oxidises bromomalonic acid, regenerating Br^- ion in the process:

$$4Ce^{4+} + BrCH(CO_2H)_2 + 2H_2O = 4Ce^{3+} + Br^- + 2CO_2 + HCO_2H + 5H^+ \tag{9.58}$$

This reaction appears to involve the participation of at least seven intermediates, and illustrates that schemes other than (9.43), due to Lotka, may give rise to sustained oscillations, provided that there is a mechanism for providing feedback. From his detailed analysis, R. M. Noyes has concluded that the important steps are as follows:

$$\left.\begin{array}{rcll} BrO_3^- + Br^- + 2H^+ & \rightleftharpoons & HBrO_2 + HOBr & (I) \\ BrO_3^- + HBrO_2 + H^+ & \rightarrow & 2BrO_2 + H_2O & (II) \\ Ce^{3+} + BrO_2 + H^+ & \rightarrow & Ce^{4+} + HBrO_2 & (III) \end{array}\right\} \qquad (9.59)$$

Combining step II with twice step III, this adds up to,

$$2Ce^{3+} + BrO_3 + HBrO_2 + 3H^+ = 2Ce^{4+} + 2HBrO_2 + H_2O \qquad (9.60)$$

which signifies the autocatalytic formation of $HBrO_2$. On this basis, Noyes and his co-workers constructed a model which has been shown to predict oscillations, particularly in the concentrations of Br^- and of Ce^{4+} . It is believed that, as a consequence of the autocatalysis in (9.60), the concentration of Br^- becomes depleted to the extent that step I is halted. This then creates the conditions for phase (iii) of the overall reaction, in which Br^- is regenerated and ceric ion is reduced back to cerous, ready for phase (i) to resume.

9.7 Catalysis by enzymes

Enzymes are proteins with a molecular weight usually in the range 10,000 to more than 100,000 and contain hundreds of amino acid units. The properties of each enzyme are a consequence of the amino acid sequence and in some cases of the metal ion (such as Fe, Co, Mn or Zn) which many contain as an essential component.

Representation of an enzyme as a simple polypeptide chain is inadequate, since its behaviour depends on the specific shape and structure, arising from interactions between the different residues of the chain. These may arise from hydrogen bonding between the C=O and N—H groups of different peptide links: where there are cysteine residues included, these are usually covalently bonded by a disulphide link.

Enzymes fulfil a substantial catalytic role in plant and animal systems. A molecule of substrate approaches an enzyme, becomes attached at some particular site on the enzyme, whereupon a reaction takes place. This must, of course, be a chemical reaction which is thermodynamically possible. A range of types of reactions is known to occur on enzymes: redox, hydrolysis, isomerisation, addition, group transfer and oxidation. The first step towards reaction involves the formation of a complex between the free molecule of substrate, S, and the enzyme, E:

$$E + S \rightleftharpoons E.S \qquad (9.61)$$

Biochemists normally use K_s to denote the equilibrium constant for the *dissociation* of E.S, the reverse of process (9.61). Typical values lie in the range 10^{-2} to 10^{-5} mol dm^{-3}, similar to those of the ionisation of a range of carboxylic acids.

In using an enzyme, almost invariably the (nominal) concentration of the enzyme will be very much less than that of the substrate. Thus, only a very small fraction of the substrate may be bound up in the enzyme–substrate complex, but for the enzyme the fraction may be quite appreciable. Suppose, for example, we add amounts of enzyme and substrate which would lead (in the absence of complex formation) to concentrations of 10^{-6} and 10^{-4} mol dm^{-3}, and that K_s has the rather lowly value of 3×10^{-5} mol dm^{-3}. We may calculate that the equilibrium concentration of E.S would be 7.7×10^{-7} mol dm^{-3}, representing 77% of the enzyme but only 0.77% of the substrate molecules present.

The simplest scheme for an enzyme catalysed reaction is that used by Michaelis and Menten in 1913. The first step is the reversible formation of the complex E.S, and this is followed by the conversion of the complexed molecule of substrate to the product and its disengagement from the enzyme:

$$\left.\begin{array}{rcl} \mathrm{E+S} & \underset{-1}{\overset{1}{\rightleftarrows}} & \mathrm{E.S} \\ \mathrm{E.S} & \overset{2}{\rightarrow} & \mathrm{E+P} \end{array}\right\} \tag{9.62}$$

It is conventional to use $[\mathrm{E}]_o$ to denote the hypothetical or nominal concentration of enzyme, calculated as the number of moles added divided by the volume of solution, and [E] for the actual concentration, allowing for complex formation. Thus:

$$[\mathrm{E}]_o = [\mathrm{E}] + [\mathrm{E.S}] \tag{9.63}$$

Applying the Principle of Stationary States to the complex E.S, we obtain:

$$\frac{\mathrm{d[E.S]}}{\mathrm{d}t} = k_1[\mathrm{E}][\mathrm{S}] - k_{-1}[\mathrm{E.S}] - k_2[\mathrm{E.S}] = 0 \tag{9.64}$$

$$\therefore [\mathrm{E.S}] = \frac{k_1[\mathrm{E}][\mathrm{S}]}{k_{-1} + k_2} \tag{9.65}$$

Substituting from equation (9.63), this leads to:

$$[\mathrm{E.S}] = \frac{[\mathrm{E}]_o[\mathrm{S}]}{\dfrac{k_{-1} + k_2}{k_1} + [\mathrm{S}]} \tag{9.65a}$$

It is convenient to use K_M, called the Michaelis constant, to denote $(k_{-1} + k_2)/k_1$. Using equation (9.65a) to evaluate the rate of the enzyme catalysed reaction, we have:

$$\begin{aligned} \text{Rate} = v &= k_2[\mathrm{E.S}] \\ &= \frac{k_2[\mathrm{E}]_o[\mathrm{S}]}{K_\mathrm{M} + [\mathrm{S}]} \end{aligned} \tag{9.66}$$

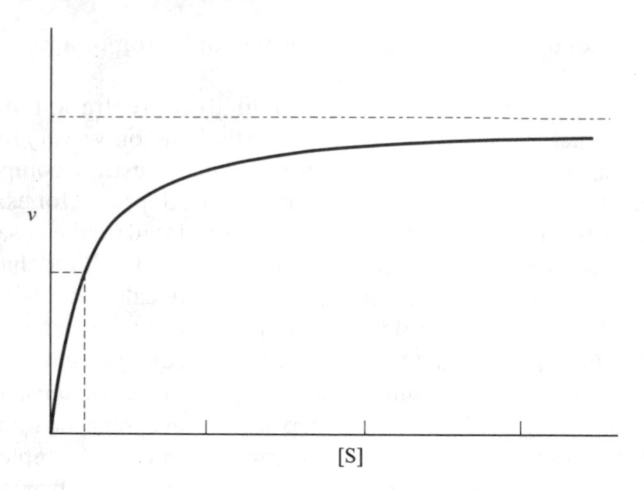

Fig. 9.7 Plot of the initial rate, v, of an enzyme-catalysed reaction as a function of the concentration of the substrate, S, based on equation (9.62). The line (-----) denotes v_{max}.

The dependence of the initial rate on the nominal concentration of substrate, given in this equation, is illustrated in Figure 9.7, which shows that, as a function of substrate concentration, the reaction rate rises sharply and then levels out to a plateau.

The shape of the curve in Figure 9.7 resembles that of the amount of gas chemisorbed on a surface as a function of the gas pressure, illustrated in Figure 8.2. Figure 9.7 means that when few enzyme molecules have been complexed, an increase in the substrate concentration causes a significant increase in the reaction rate, but when nearly all enzyme molecules are bound in complexes, then a further increase has little effect on the reaction rate.

By taking the reciprocal of each side of equation (9.66) it becomes,

$$\begin{aligned} \frac{1}{v} &= \frac{\mathrm{K_M}}{k_2[\mathrm{E}]_o[\mathrm{S}]} + \frac{1}{k_2[\mathrm{E}]_o} \\ &= \frac{1}{k_2[\mathrm{E}]_o}\left(\frac{K_\mathrm{M}}{[\mathrm{S}]} + 1\right) \end{aligned} \tag{9.67}$$

which shows that when v^{-1} is plotted against $[\mathrm{S}]^{-1}$, from the intercept and slope of the resulting straight line it is possible to derive k_2 and K_M. The latter is equal to the value of the concentration of S at which the reaction rate is equal to half of the plateau value, and is indicated by the dotted line in Figure 9.7.

Referring to the reaction steps in mechanism (9.62), in many cases $k_{-1} \gg k_2$. Consequently, the Michaelis constant K_M is effectively equal to K_s, the dissociation constant for the enzyme–substrate complex. In general,

$K_M \leq K_s$, so that the experimental value for the Michaelis constant can serve as a minimum value for K_s.

While a linear plot of v^{-1} against $[S]^{-1}$ is predicted by equation (9.67), it does not necessarily mean that the behaviour of the substrate on the enzyme conforms exactly to mechanism (9.62). The point is that a more complex model may lead to a very similar equation. For example, suppose the substrate S reacts on the enzyme to yield two products, X and Y, produced by the following scheme,

$$\left.\begin{array}{rcl} E + S & \underset{-1}{\overset{1}{\rightleftarrows}} & E.S \\ E.S & \overset{2}{\rightarrow} & E.S' + X \\ E.S' & \overset{3}{\rightarrow} & E + Y \end{array}\right\} \qquad (9.68)$$

in which X and Y are released from the enzyme sequentially. This leads to the rate expression,

$$v = \frac{k_2\chi[E]_o[S]}{K_M\chi + [S]} \qquad (9.69)$$

where $\chi = k_3/(k_2 + k_3)$. This differs only very slightly from equation (9.66).

9.8 Enzyme inhibitors

The ability of an enzyme to act as a catalyst is dependent on the active site being available to accept a molecule of substrate, so that the enzyme can then act on it. Substances which interfere with the active sites are called enzyme inhibitors.

A foreign substance may interact very strongly with an enzyme, bonding covalently to it close to the active site and thus destroying the enzyme activity. For example, diisopropyl fluorophosphate attacks acetylcholinesterase and renders it inactive. Since acetylcholine is an essential neurotransmitter, the effective removal of this enzyme causes serious dysfunction in mammalian systems. Such is the nature of chemical warfare, using nerve gases to which the body is extremely vulnerable.

Other substances which bind to the enzyme in a similar way to the substrate function as reversible inhibitors, capable of being displaced by substrate molecules leaving the enzyme intact. We will consider one such model in more detail, by adding to mechanism (9.62) the reversible step,

$$E + I \rightleftharpoons E.I \qquad (9.70)$$

where I represents the inhibitor and E.I the enzyme–inhibitor complex. We will denote the dissociation constant of E.I by K_I, so that:

$$[E.I] = [E][I]/K_I \qquad (9.71)$$

In the presence of the inhibitor, equation (9.63) gives way to the relation,

$$[E]_o = [E] + [E.S] + [E.I] = [E.S] + [E]\{1 + [I]/K_I\} \tag{9.72}$$

The Stationary State equation for E.S is unaffected by the presence of the inhibitor, so that equation (9.65) still applies, and we may write it as:

$$[E] = [E.S]\, K_M/[S] \tag{9.65a}$$

Substituting this into equation (9.72), and rearranging, we have:

$$[E.S] = \frac{[E]_o}{1 + \left\{1 + \frac{[I]}{K_I}\right\} \frac{K_M}{[S]}} \tag{9.73}$$

Now that we have an expression for E.S concentration in terms of the concentrations of substrate and of inhibitor and of the nominal concentration of enzyme, along with K_M and K_I, we can obtain a useful expression for the rate of the inhibited enzyme-catalysed reaction.

$$\text{Rate} = v = k_2[E.S] = \frac{k_2[E]_o[S]}{[S] + K_M\{1 + [I]/K_I\}} \tag{9.74}$$

The value of this expression will always be less than that of equation (9.66) for the uninhibited reaction, but both tend towards the same upper limit at high concentrations of substrate. If the concentration of I is equal to K_I, then at a substrate concentration equal to K_M, the rate is now one-third of the maximum achievable value.

To measure the equilibrium constant K_I, data are required on the initial rate of the enzyme-catalysed reaction as a function of substrate concentration, at a fixed concentration of the inhibitor. Inverting equation (9.74) we have:

$$\frac{1}{v} = \frac{1}{k_2[E]_o} + \frac{K_M}{k_2[E]_o[S]} \left\{1 + \frac{[I]}{K_I}\right\} \tag{9.75}$$

This shows that at constant inhibitor concentration, v^{-1} is a linear function of $[S]^{-1}$ and the ratio of slope to intercept yields $K_M\{1 + [I]/K_I\}$. The relation between this plot and that for the uninhibited reaction is illustrated in Figure 9.8, which shows that both have the same intercept on the ordinate axis.

The type of molecule which can act as an inhibitor is one which is structurally similar to the substrate, such as a homologue. For example, malonic acid is complexed but not acted upon by succinate dehydrogenase and thus acts as an inhibitor.

Our model of a molecule complexing with the enzyme but not then undergoing reaction may be used in regard to the stereoselective aspect of enzyme action. All enzymes are made up from amino acid residues which, glycine apart, are optically active, so that each enzyme possesses chirality. In consequence, an enzyme catalyses the reaction of one particular enantiomer, but not the other and not a diastereomer.

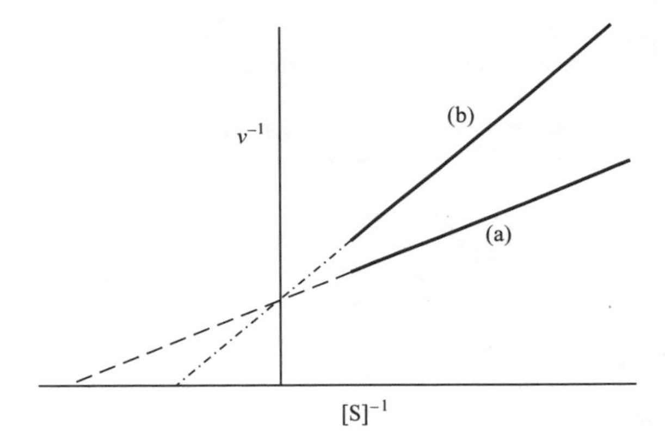

Fig. 9.8 Notional plots of v^{-1} against $[S]^{-1}$ (the so-called "Lineweaver–Burk plot") for (a) an uninhibited reaction, which yields K_M, and (b) a reaction in the presence of a constant concentration of inhibitor, which yields K_I when K_M is known.

In many cases, the "wrong" enantiomer binds at the active site of the enzyme just as strongly as does the correct one, but since it has the wrong stereochemistry the enzyme is unable to act upon it so that $k_2 = 0$. Thus the ethyl ester of D-tyrosine is complexed to chymotrypsin just as strongly as is that of the naturally occurring L-tyrosine, but no reaction takes place. In this instance, the "wrong" enantiomer behaves as an enzyme inhibitor.

Suggested reading

Leisten, J. A., 1964. Homogeneous catalysis: a re-examination of definitions. *Journal of Chemical Education*, **41**, 23–7.

Gates, B. C., 1992. *Catalytic Chemistry*. New York: Wiley.

Bender, M. L. and L. J. Brubacher, 1973. *Catalysis and Enzyme Action*. New York: McGraw-Hill.

Isaacs, N. S., 1987. *Physical Organic Chemistry*. London: Longman.

Maskill, H., 1985. *The Physical Basis of Organic Chemistry*. Oxford: OUP.

Piszkiewicz, D., 1977. *Kinetics of Chemical and Enzyme-catalyzed Reactions*. New York: OUP.

Mata-Perez, F. and J. F. Perez-Benito, 1987. The kinetic rate law for autocatalytic reactions. *Journal of Chemical Education*, **64**, 925–7.

Physical chemistry of oscillatory phenomena, 1974. *Symposia of the Faraday Division*, No. 9. London: The Chemical Society.

Degn, H., 1972. Oscillating chemical reactions in homogeneous phase. *Journal of Chemical Education*, **49**, 302–7.

Field, R. J., 1972. A reaction periodic in time and space. *Journal of Chemical Education*, **49**, 308–11.

Field, R. J. and R. M. Noyes, 1974. Oscillations in chemical systems. IV Limit cycle behaviour in a model of a real chemical reaction. *Journal of Chemical Physics*, **60**, 1877–84.

Lotka, A. J., 1920. Undamped oscillations derived from the law of mass action. *Journal of the American Chemical Society.*, **44**, 1595–9.

Problems

9.1 Suppose that for a halogen other than Br which we shall call X, the reaction steps in the presence of ozone (mechanism (9.2)) are slightly altered in that step 4 is replaced by:

$$X + X \xrightarrow{5} X_2$$

(a) What then is the rate expression for the formation of O_2?
(b) Why does this reaction show a dependence on X_2 concentration which differs from the rate dependence on Br_2 shown in equation (9.7)?
(c) Is X_2 a catalyst for the decomposition of ozone?

9.2 In 3.0 mol dm^{-3} perchloric acid with traces of silver ion present, Co^{III} can oxidise Cr^{III} to Cr^{VI} ion. In runs at 298 K in which Cr^{III} concentration was 1.25 x 10^{-2} mol dm^{-3}, Co^{II} concentration was 0.012 mol dm^{-3} and that of Co^{III} was initially 4×10^{-3} mol dm^{-3}, the formation of Cr^{VI}, monitored by the rising absorbance at 475 nm, was found to follow first order kinetics. From the plots of $\ln(A_\infty - A_t)$ against time, the first order rate constant was evaluated and found to be directly proportional to the concentration of Ag^{I} over the range, 4×10^{-5} to 4×10^{-3} mol dm^{-3}.

From runs at 298 K in which the concentration of Ag^{I} was 1.9×10^{-4} mol dm^{-3}, and those of Co^{II} and Co^{III} were initially 0.005 and 0.014 mol dm^{-3}, the first order rate constant, k_{obs}, was found to vary with the Cr^{III} concentration as follows:

$[Cr^{III}]/10^{-3}$ mol dm^{-3}	$k_{obs}/10^{-3}\,s^{-1}$
126	4.44
41	3.18
32	3.02
25	2.62
12.5	1.90
6.2	1.01

Suggest a possible mechanism of this reaction and show that it is consistent with these data. (Data from Kirwin, J. B., P. J. Proll and L. H. Sutcliffe, *Trans Faraday Soc.*, 1964, **60**, 119–26.)

9.3 From experiments on the hydrolysis of *p*-nitrophenyl acetate in 0.2 mol dm^{-3} phosphate buffer at 30°C, first order plots were made by plotting $\log_{10}\{a/(a - x)\}$ against time. The slopes, k', of these plots as a function of the concentration of added imidazole, **Im**, vary as follows:

Im/mol dm^{-3}	k'/ min^{-1}
0	1.91×10^{-3}
0.5×10^{-4}	2.34×10^{-3}
1.0×10^{-4}	2.72×10^{-3}
1.5×10^{-4}	3.16×10^{-3}
2.0×10^{-4}	3.64×10^{-3}

If the pseudo-first order rate constant, k' is given by: $k' = k_o + k_I[I]$, deduce the values of k_o and k_I. (Data from Bruice, T. C. and G. L. Schmir, *J. Am. Chem. Soc.*, 1957, **79**, 1663–7.)

9.4 With a constant concentration of glutamate dehydrogenase, the initial rate of breakdown, v, of glutamic acid is found to vary with the substrate concentration as follows:

[S]/10^{-2} mol dm^{-3}	0.16	0.25	0.40	0.70	1.00	1.50	2.00
v/10^{-5} mol dm^{-3} s^{-1}	2.23	2.82	3.48	4.17	4.45	4.76	4.96

Assuming that the Michaelis–Menten equation is applicable, evaluate the maximum initial rate at high substrate concentrations and the Michaelis constant, K_M.

9.5 The hydration of fumarate to maleate by the enzyme fumarase is inhibited by succinate. Initial rates at several concentrations of fumarate, in the absence and the presence of 0.05 mol dm^{-3} succinate, are given below.

[S]/10^{-4} mol dm^{-3}	No inhibitor v/arbitrary units	With 0.05 mol dm^{-3} succinate present v/arbitrary units
5.0	3.36	2.74
1.5	2.45	1.72
0.8	1.79	1.13
0.5	1.33	0.79

Determine whether the action of succinate on this enzyme fits that of competitive inhibition and if so determine the value of K_I for succinate on fumarase.

10 Relaxation and other advanced techniques

The problems inherent in studying the kinetics of fast reactions in solution by mixing solutions of the reactants have been referred to in earlier chapters. One approach uses the fact that, in a homogeneous phase, no reaction attains completion at equilibrium. A group of techniques uses the re-establishment of equilibrium after the solution has been acted upon so that the position of equilibrium has been suddenly perturbed. In this way a rapid reaction may be studied kinetically without the need to solve the dilemma of rapid but thorough mixing.

Where the equilibrium in solution is a single reversible process, such data are easily interpreted. In some cases, where there are a number of interlinked equilibria, the interpretation of the relaxation data is much less straightforward. Other techniques appropriate to fast reactions in solution are also considered, including applications of nuclear magnetic resonance (NMR) spectroscopy and of polarography.

10.1 Introduction

In studying reactions in solution, a major limitation arises because of the vital importance of thorough mixing and the time required (at least *ca* 10^{-3} seconds) to achieve this. If the rate constants of the reactions to be studied are high then, assuming these to be second order reactions, they can be studied in an accessible time scale only by using very low concentrations. However, this may make reaction monitoring very difficult. If spectrophotometry is used, it may mean extremely small changes in the absorbance or else cells of long path length.

One way to solve the problem is to produce *in situ* one or both of the species that will engage in the reaction. Photochemical and radiation chemical methods can be used to produce transient species so that their reactions may be studied. There are, however, many reactions for which these techniques are inapplicable.

A group of methods appropriate to such conditions was developed around 1950 and, for his contribution to these endeavours, Eigen won a share of the Nobel Prize for Chemistry in 1967. These methods obviate the need for mixing reactants by achieving a system which is slightly perturbed from its equilibrium position. A technique such as kinetic spectrophotometry or conductivity is then used to monitor the approach to equilibrium as the chemical system undergoes *relaxation*.

The means of obtaining a slightly perturbed system may be illustrated by reference to the temperature jump technique. Only rarely is the standard enthalpy change of a chemical reaction effectively zero. In the light of the Van't Hoff equation,

$$\ln K = -\frac{\Delta H^{\ominus}}{RT} + \frac{\Delta S^{\ominus}}{R} \tag{10.1}$$

a change of temperature must entail some change in the equilibrium constant, K. This means that if the temperature of a system which was in equilibrium can suddenly be raised by a few degrees, the system becomes perturbed and will then relax towards a new position of equilibrium.

In kinetic terms, the simplest system which might undergo relaxation has first order (or pseudo-first order) reactions, both forward and reverse:

$$\mathrm{A} \underset{-1}{\overset{1}{\rightleftarrows}} \mathrm{B} \tag{10.2}$$

Let us denote the equilibrium concentrations by a and b and the actual concentrations at time t by $(a + x)$ and $(b - x)$. The rate equation is then:

$$\begin{aligned} -\frac{\mathrm{d}x}{\mathrm{d}t} &= k_1(a+x) - k_{-1}(b-x) \\ &= k_1 a - k_{-1} b + (k_1 + k_{-1})x \\ &= (k_1 + k_{-1})x \end{aligned} \tag{10.3}$$

To attain the last line, one uses the fact that $k_1 a = k_{-1} b$, since these are the equilibrium concentrations. Separating the variables and integrating, this leads to:

$$x = x_0 \mathrm{e}^{-(k_1 + k_{-1})t} \tag{10.4}$$

This means that the extent of the perturbation decreases exponentially, with the pseudo-first order rate constant for its disappearance equal to the sum of the forward and reverse rate constants (see equation 3.57). In studying physical phenomena, it is normal to use the term, relaxation time, τ, to denote the interval needed for the extent of the perturbation to diminish by the factor, $1/e$. So for this system, the relaxation time is given by:

$$\tau = (k_1 + k_{-1})^{-1} \tag{10.5}$$

Alternatively,

$$\tau^{-1} = k_1 + k_{-1} \tag{10.5a}$$

Combined with a knowledge of the equilibrium constant, this enables k_1 and k_{-1} to be evaluated.

Let us now consider the reaction system in which one reaction is second order and the opposing one is first order:

$$\mathrm{A + B} \underset{-1}{\overset{1}{\rightleftarrows}} \mathrm{C} \tag{10.6}$$

If the equilibrium concentrations are denoted by *a*, *b* and *c*, then the rate equation is

$$\begin{aligned} -\frac{\mathrm{d}x}{\mathrm{d}t} &= k_1(a+x)(b+x) - k_{-1}(c-x) \\ &= k_1ab + k_1(a+b)x + k_1x^2 - k_{-1}c + k_{-1}x \\ &= x\{k_1(a+b) + k_{-1}\} \end{aligned} \tag{10.7}$$

In this case, in addition to using our knowledge of the equilibrium concentrations, we have assumed that the perturbation is slight so that the term in x^2 is negligible. Thus for reaction (10.6) we have:

$$\tau^{-1} = k_1(a+b) + k_{-1} \tag{10.8}$$

In order to determine k_1 and k_{-1} it is now necessary to vary the equilibrium concentrations of A and B. A plot of τ^{-1} against $(a + b)$ has a slope of k_1 and an intercept of k_{-1}.

For a system in which both the forward and the reverse reactions are second order,

$$\mathrm{A + B} \underset{-1}{\overset{1}{\rightleftarrows}} \mathrm{C + D} \tag{10.9}$$

the rate equation leads, in similar fashion, to the relation:

$$\tau^{-1} = k_1(a+b) + k_{-1}(c+d) \tag{10.10}$$

However, if $k_1 = k_{-1}$, then the validity of equation (10.10) is not restricted to the regime of small perturbations, but is generally valid.

Example 10.1

In a laser temperature jump study of the equilibrium,

$$\mathrm{I^- + I_2} \underset{-1}{\overset{1}{\rightleftarrows}} \mathrm{I_3^-}$$

the relaxation time, τ, at 298 K was found to vary as follows with the equilibrium concentrations of iodide and iodine:

$[I^-]/10^{-3}$ mol dm^{-3}	$[I_2]/10^{-3}$ mol dm^{-3}	τ/ns
0.57	0.36	70.7 ± 3.5
1.58	0.24	50.0 ± 5.2
2.39	0.39	39.0 ± 4.4
2.68	0.16	37.9 ± 4.4
3.45	0.14	32.4 ± 4.0

Thus evaluate the forward and reverse rate constants, k_1 and k_{-1}. (Data from Turner, D. H., G. W. Flynn, N. Sutin and J. V. Beitz, *J. Am. Chem. Soc.*, 1972, **94**, 1554–9.)

Answer. The relevant equation is:

$$\tau^{-1} = k_1\{[I^-] + [I_2]\} + k_{-1} \tag{10.8}$$

so it is appropriate to plot τ^{-1} against $\{[I^-] + [I_2]\}$. Thus we have:

$\{[I^-] + [I_2]\}/10^{-3}$ mol dm^{-3}	$\tau^{-1}/10^7$ s^{-1}
0.93	1.414 ± 0.07
1.82	2.00 ± 0.21
2.78	2.56 ± 0.30
2.84	2.64 ± 0.31
3.59	3.09 ± 0.38

From the plot of τ^{-1} against $\{[I^-] + [I_2]\}$ in Figure 10.1, it may be seen that a good straight line is obtained. From the slope and the intercept of this line, k_1 is determined as 6.3×10^9 dm^3 mol^{-1} s^{-1} and k_{-1} as 8.3×10^6 s^{-1}.

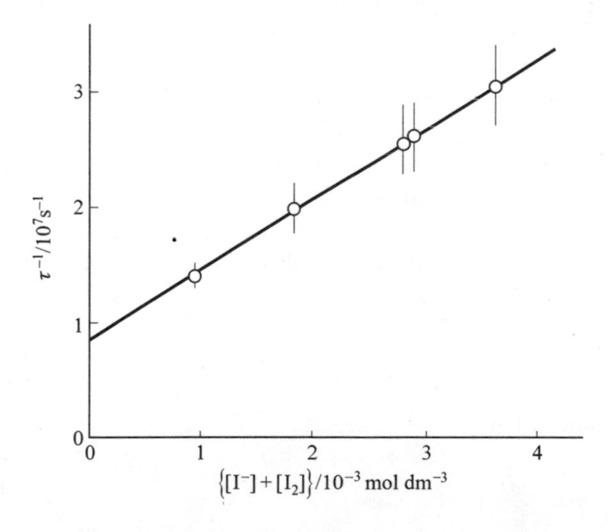

Fig. 10.1 Plot of τ^{-1} against $\{[I^-] + [I_2]\}$ for temperature jump experiments on the system, $I^- + I_2 \rightleftarrows I^-_3$, using the data listed.

An alternative approach is to use the fact that $k_1/k_{-1} = K_1$, the known equilibrium constant.

Thus equation (10.8) may be re-written,

$$k_1 = \tau^{-1}\{[I_2] + [I^-] + K_1^{-1}\}^{-1}$$

so that the forward rate constant may be evaluated from each determination.

τ/ns	$k_1/10^9\ dm^3\ mol^{-1}\ s^{-1}$
70.7 ± 3.5	6.15 ± 0.3
50.0 ± 5.2	6.27 ± 0.6
39.0 ± 4.4	6.18 ± 0.6
37.9 ± 4.4	6.27 ± 0.7
32.4 ± 4.0	6.22 ± 0.7

So, $k_1 = (6.2 \pm 0.7) \times 10^9\ dm^3\ mol^{-1}s^{-1}$.

10.2 Relaxation techniques: experimental aspects

The most widely used relaxation method is the temperature jump technique, whose rationale was mentioned above. Early versions involved the use of joule heating on a small sample of solution, positioned between two optical faces so that the resulting reaction could readily be monitored by kinetic spectrophotometry, as illustrated in Figure 10.2.

Rapid joule heating may be achieved by the rapid discharge, through the solution, of a high voltage capacitor. However, it is necessary for the solution to contain a moderate concentration of electrolyte to boost the electrical conductivity. This effectively limits the technique to solvents of high relative permittivity. A temperature rise of several degrees (even up to 10 K) may be attained, depending on the experimental details. For example, a capacitor of 0.05 μF charged to 30 kV has a stored energy of $\frac{1}{2}CV^2 = 22.5$ J. This may be expended in less than 1 cm^3 of solution: if it were 1 cm^3 of water then the temperature should rise by *ca* 5.4 K.

The extent of the reaction occasioned by such a jump in temperature depends on the value of the standard enthalpy change. Taking the Van't Hoff equation, (10.1) in the form,

$$\frac{\mathrm{d}\ln K}{\mathrm{d}(1/T)} = \frac{-\Delta H^{\ominus}}{R} \qquad (10.11)$$

this leads to the relation:

$$\frac{\mathrm{d}K}{K} = \frac{\Delta H^{\ominus}}{R} \cdot \frac{\mathrm{d}T}{T^2} \qquad (10.12)$$

If $\Delta H^{\ominus} = 20$ kJ mol^{-1}, then for a temperature rise of 5 K, at 298 K there is a change of 13.3% in the value of the equilibrium constant.

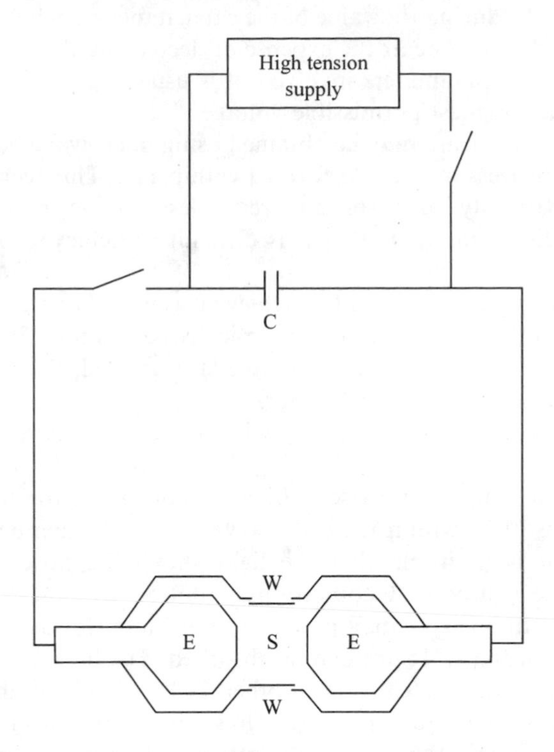

Fig. 10.2 A sketch of a temperature-jump cell employing joule heating of the solution. C, capacitor; E, electrodes; S, sample; W, windows for the (monochromatic) analysing light.

However, the consequence of such a temperature rise depends also on the value of the equilibrium constant and on the concentration of the species involved. In general, the greatest relative change in the concentration is achieved where the concentrations of the reactant and product species are comparable. But the technique by which the re-equilibration is to be monitored may itself impose some constraints on these concentrations.

Another important parameter is the rapidity of the capacitor discharge. The current falls off exponentially, so that the final temperature is approached asymptotically, according to the equation,

$$\Delta T_t = \Delta T_\infty(1 - \mathrm{e}^{-t/\tau}) \qquad (10.13)$$

where ΔT_t denotes the temperature rise attained at time t, ΔT_∞ the maximum rise and $\tau = RC/2$, where R is the resistance of the discharge circuit and C the value of the capacitance. Obviously the resistance of the solution should be kept low.

In the example quoted earlier, the capacitance was $0.05\,\mu$F. If R were $250\,\Omega$, then $\tau = 6.25\,\mu$s, which means that 95% of the increase in temperature would occur in 19 μs. Reducing the value of the capacitance may help to achieve a faster rise, but this may be at the expense of decreasing the extent of the rise. In optimising the parameters involved, it is usually desirable to charge the capacitor to the highest permissible voltage.

More rapid jumps in temperature may be obtained using microwave heating, where the complete process may be performed within 1 μs. This technique is applicable to virtually any solvent, regardless of its relative permittivity. The main disadvantage is that it is difficult to achieve a rise in excess of 1 K.

The most rapid temperature jump of all is achieved using a laser pulse, where the pulse width and thus the rise time will typically be less than 30 ns. In one such design, light of 1.315 μm from an iodine laser is used. Whereas at the wavelength of the neodymium YAG laser, 1.06 μm, water is almost transparent, at 1.315 μm it absorbs moderately with a decadic absorption coefficient of $0.76\,\mathrm{cm}^{-1}$.

To achieve a significant temperature rise, a high proportion of the incident light needs to be absorbed within the cell. However, for this temperature rise to be useful, it is desirable that the light should be absorbed uniformly throughout the solution. A compromise solution to these twin problems must be sought. In one system, a path length of 3 mm was used, in the course of which 41% of the 1.315 μm light is absorbed. The light is then reflected back by a plane mirror, causing another 24% of the original intensity to be absorbed on the return passage. Thus almost two-thirds of the laser pulse is absorbed within the cell and the rate of light absorption at the back of the cell is only 13% less than at the front. This modest extent of deviation from complete homogeneity is acceptable, particularly since the reactions being observed are exhibiting pseudo-first order kinetics.

Another relaxation method is that of pressure jump, which relies on the fact that the standard molar volume change for a reaction is usually non-zero:

$$\frac{\mathrm{d}\ln K}{\mathrm{d}P} = \frac{-\Delta V^{\ominus}}{RT} \tag{10.14}$$

However, only for a few reactions is $\Delta V^{\ominus}$ sufficiently large to suggest that the method is specially appropriate.

The pressure jump is almost invariably achieved in reverse, starting with the solution under a pressure of perhaps 50 atmospheres, and releasing this pressure by puncturing a thin metal diaphragm. The time required for such release is *ca* 50 μs, so the technique can not match the time resolution of a good temperature jump system.

The third widely used relaxation method is the dissociation field effect, in which a large electric field is applied to the solution to perturb ionic equilibria. To minimise the heating effect, the concentration of the electrolyte should be kept low and the pulse should be applied only for *ca* 20 μs. The equilibrium is then sufficiently perturbed and the relaxation is best followed by a conductimetric method.

The dissociation field effect was applied in the 1950s to the reaction:

$$H^+ + OH^- \underset{-1}{\overset{1}{\rightleftarrows}} H_2O \tag{10.15}$$

The relaxation time of this system was measured as 37 μs, from which, using the ionic product $K_w = 1.0 \times 10^{-14}\,\text{mol}^2\,\text{dm}^{-6}$ and equation (10.7), k_1 is readily evaluated as $1.35 \times 10^{11}\ \text{dm}^3\,\text{mol}^{-1}\,\text{s}^{-1}$.

10.3 More complex reaction systems

Chemical systems actually studied are often much more complex than those considered in Section 10.1, and so may require rather more ingenuity in their interpretation. As an example, let us consider the coupled equilibria:

$$A + B \rightleftarrows C \rightleftarrows D \tag{10.16}$$

This mechanism may be broken down into two obvious parts. For the first of these,

$$A + B \underset{-1}{\overset{1}{\rightleftarrows}} C \qquad \textbf{(I)}$$

we have the relation,

$$\tau_1^{-1} = k_1(a + b) + k_{-1} \tag{10.17}$$

and for the second,

$$C \underset{-2}{\overset{2}{\rightleftarrows}} D \qquad \textbf{(II)}$$

the relaxation time is given by:

$$\tau^{-1} = k_2 + k_{-2} \tag{10.18}$$

However, the fact that the two equilibria are coupled has an appreciable effect on the nature of these relaxation processes.

In some cases one equilibrium is established considerably more quickly than the other: let us suppose this is equilibrium **(I)**. Then the system will exhibit a relaxation time very close to that specified in equation (10.17). However, it is important to appreciate that in the perturbation, the concentrations of A, B, C and D were all displaced from their equilibrium values. The rapid relaxation to which we are referring merely brings the concentrations of A, B and C towards values which satisfy the equilibrium constant K_1. However, the concentration of D has not yet attained its final value, so the concentrations attained cannot be the final equilibrium values of these three species. Thus after the first relaxation process is essentially complete, the process that remains is rather more complex than reaction **(II)**.

Let us suppose that the equilibrium concentrations of the species involved are, respectively, a, b, c and d. Further, let us suppose that, after the relaxation of step (I), the actual concentrations are $(a - x)$, $(b - x)$, $(c - y)$ and $(d + z)$. Two consequences follow:

(i) Since equilibrium (I) is established, we have:

$$K_1 = \frac{c - y}{(a - x)(b - x)} = \frac{c}{ab} \tag{10.19}$$

Ignoring the term in x^2, y can now be expressed in terms of x:

$$1 - \frac{y}{c} = 1 - \frac{(a + b)x}{ab} \tag{10.20}$$

$$\therefore y = K_1(a + b)x \tag{10.21}$$

(ii) From considerations of material balance, $(a - x)$, $(c - y)$ and $(d + z)$ must add up to $(a + c + d)$, so:

$$z = x + y \tag{10.22}$$

Using equations (10.21) and (10.22), y may be expressed in terms of z:

$$y\left\{1 + \frac{1}{K_1(a + b)}\right\} = z \tag{10.23}$$

$$\therefore y = \frac{K_1(a + b)z}{K_1(a + b) + 1} \tag{10.24}$$

Since D is the only species which is involved solely in the slow relaxation process, the natural variable to use here is z:

$$\begin{aligned}\frac{dz}{dt} &= k_2(c - y) - k_{-2}(d + z) \\ &= k_2\left\{c - \frac{K_1(a + b)z}{K_1(a + b) + 1}\right\} - k_{-2}(d + z) \\ &= -\left\{k_2 \frac{K_1(a + b)}{K_1(a + b) + 1} + k_{-2}\right\}z\end{aligned} \tag{10.25}$$

Thus, by analogy with equation (10.3), the second relaxation time is given by:

$$\tau_2^{-1} = k_2 \frac{K_1(a + b)}{K_1(a + b) + 1} + k_{-2} \tag{10.26}$$

If equilibrium (I) lies well to the right (as will happen at moderate concentrations if K_1 is large), then equation (10.26) scarcely differs from equation (10.18). However, where the product, $K_1(a + b)$ is either around unity or much less than unity, then the reciprocal of the relaxation time becomes less than that for the uncoupled step (II) on its own. In the limit, for $K_1(a + b) \ll 1$, equation (10.26) tends towards k_{-2}.

Evaluating k_2 and k_{-2} from the dependence of τ_2 on the concentrations a and b, using equation (10.26), is not always feasible. The plot of τ_2^{-1} against

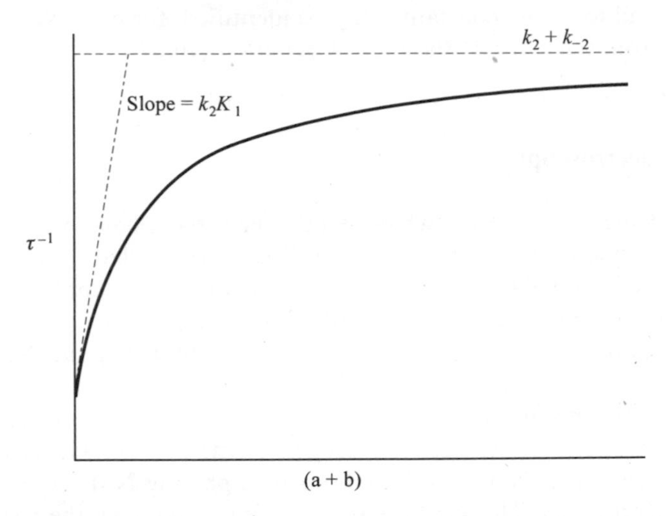

Fig. 10.3 Plot of τ^{-1} against $(a+b)$ from equation (10.26), using selected values of k_2 and k_{-2} .

$(a+b)$, sketched in Figure 10.3, is curved. Correctly extrapolated, it has an intercept of k_{-2} and a limiting slope of k_2K_1. At high values of the ordinate, it attains a plateau value of $(k_2 + k_{-2})$. However, depending on the value of K_1, in many instances it is not feasible to use this curve accurately to derive both the forward and the reverse rate constants for step **(II)**.

The foregoing discussion assumes that equilibrium **(I)** relaxes much faster than **(II)**. An alternative possibility is that the rates of these two processes are comparable. Consequently, the approximations made above are no longer valid and the equations for the relaxation of this system become more complex. In general, the system will exhibit two relaxation times, but because of the mutual coupling of the two equilibria, neither relaxation time is so simple to interpret as when one process is much faster than the other.

The appropriate mathematical analysis of this situation leads to the equations:

$$\frac{1}{\tau_1} + \frac{1}{\tau_2} = k_1(a+b) + k_{-1} + k_2 + k_{-2} \tag{10.27}$$

$$\frac{1}{\tau_1\tau_2} = k_1(k_2 + k_{-2})(a+b) + k_{-1}k_{-2} \tag{10.28}$$

To identify the individual rate constants involved, the two relaxation times should be measured as functions of $(a+b)$. As equations (10.27) and (10.28) show, both the sum and the product of the reciprocal relaxation times are linear functions of this parameter. From the intercepts and the slopes of these two plots, all four rate constants may be identified: for example, k_1 is found directly from the slope of the plot of equation (10.27).

10.4 NMR spectroscopy

After the fundamental research which established the phenomenon of nuclear magnetic resonance, a seminal paper in 1951 showed that the proton NMR spectrum of ethanol consisted of three groups of peaks, corresponding to the methyl, the methylene and the hydroxyl protons of that molecule. This served to announce the arrival of a new and powerful technique, NMR spectroscopy, for the determination of molecular structure.

Shortly after this, another important paper dealt with the effect on the NMR spectrum of the interchange of hydrogen atoms between chemically non-equivalent sites. This laid the foundations for applying NMR spectroscopy to kinetic processes. The aim here is to present, in context, the results of this analysis in a simplified and readily intelligible form.

A nuclide with a non-zero nuclear spin generates a magnetic field and thus has a nuclear magnetic moment. When such a nucleus, with a nuclear spin $= I$, is subjected to an external magnetic field, its alignment with respect to the field depends on the magnetic quantum number, m, which can adopt any value within the range, I, $(I-1)$, $(I-2)$, . . . $(1-I)$, $-I$. Thus the number of possible values for m is equal to $(2I+1)$. If $I = ½$, as it is for the hydrogen atom, then there are the two values, $m = +½$ and $m = -½$. The extent of the separation of the energy of these states is proportional to the strength, $\boldsymbol{B}_o$, of the magnetic induction field, as is illustrated in Figure 10.4.

In accordance with the Boltzmann distribution, the population of the state with $m = +½$ is very slightly greater than that with $m = -½$. A transition between these states is possible, provided that a quantum of energy of the correct size is available. However, the frequency of the quantum required to effect the transition from $m = +½$ to $m = -½$ is not constant for all protons within the sample, but depends on their chemical environment. This variation, referred to above in regard to the early study on ethanol, is known as the chemical shift and is a consequence of the differing extents to which the magnetic field experienced by the nucleus of a hydrogen atom is modified as a consequence of the shielding by the surrounding electron cloud.

On the basis of the above, if the magnetic induction field $\boldsymbol{B}_o$ were totally uniform throughout the sample being studied, then, for protons in one particular chemical environment the frequency ν of the radiation required to effect the transition would have zero range. In practice, NMR peaks have a finite width, arising, not from inadequacies in the equipment, but from the Heisenberg Uncertainty Principle:

$$\Delta\varepsilon.\Delta t \approx h/2\pi \qquad (10.29)$$

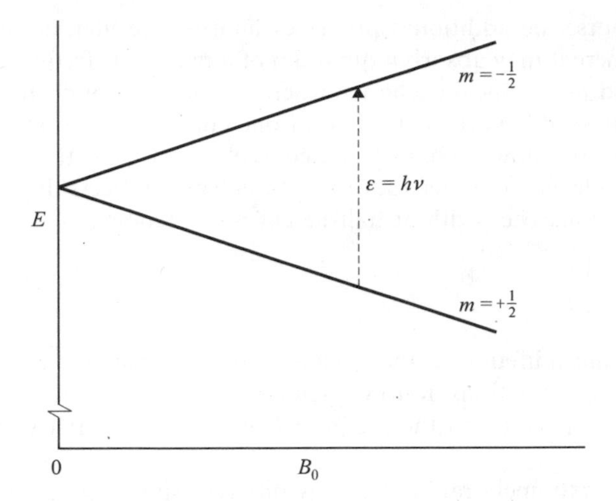

Fig. 10.4 Illustration of the separation of the energies of protons with magnetic quantum numbers +½ and −½ as a function of the magnetic field, $\boldsymbol{B}_o$.

where $\Delta\varepsilon$ and Δt represent the uncertainties in the energy and the time and h is Planck's constant. Since the uncertainty in the energy is proportional to that in the resonance frequency,

$$\Delta\varepsilon = h.\Delta\nu \tag{10.30}$$

it follows that the uncertainty in the frequency is given by:

$$\Delta\nu = (2\pi\Delta t)^{-1} \tag{10.31}$$

Apart from the transitions effected by the absorption of resonance radiation, the lifetime of a nucleus in a state with a particular value of m cannot be infinite. As nuclei move relative to each other, there is inevitable interaction of their magnetic dipoles, which allows energy transfer between the nuclear spin system and molecular motions. This constitutes a mechanism by which the disruption of the Boltzmann distribution by resonance absorption is counteracted. The populations of the various magnetic quantum numbers approach equilibrium (i.e. they *relax*) in a first order manner. The reciprocal of the rate constant is equal to the mean lifetime of a nucleus in a particular spin state (see equation 1.34) and is called the relaxation time. The relevant parameter here is known as the spin–spin or the transverse relaxation time, T_2. Thus we have, for the peak width at half-height:

$$\Delta\nu = \frac{1}{2\pi T_2} \tag{10.32}$$

In many systems, the peaks are sufficiently narrow that it may be deduced that the value of T_2 must be at least one second.

There may, of course, be additional processes limiting the lifetime of a nucleus in a state where it may absorb a quantum of a particular frequency. One such is provided by chemical exchange where, on account of chemical reaction, the nucleus spends part of its time in one chemical environment and part in another. For slow exchange, the consequence is that the peak becomes further broadened, and this additional broadening reflects the rate of the exchange reaction. The width at half-height now becomes,

$$\Delta\nu = \frac{1}{2\pi}\left(\frac{1}{T_2} + \frac{1}{\tau}\right) \tag{10.33}$$

where τ denotes the mean lifetime of the nucleus in that particular chemical environment. Exchange reactions follow pseudo-first order kinetics (see Section 3.8), so that τ is equal to the reciprocal of the pseudo-first order rate constant.

Where exchange is extremely rapid, a totally different situation prevails. Let us suppose that, with the value of $\boldsymbol{B}_o$ of the order of 1 tesla, in two different chemical environments, proton resonances occur at the radio frequencies ν_A and ν_B, and further that the amounts of the protons in these two chemical environments have mole fractions X_A and X_B. Thus, provided that τ_A and τ_B, the lifetimes of the proton in these respective environments, are both much less than $(\nu_A - \nu_B)^{-1}$, the inverse of the separation of these frequencies, a radiofrequency quantum is being absorbed by a nucleus whose time-averaged chemical environment will respond to the frequency:

$$\nu' = X_A\nu_A + X_B\nu_B \tag{10.34}$$

Further, the line width will be characteristic of an averaged spin–spin relaxation time, $T_2{}'$, given by

$$\frac{1}{T_2'} = \frac{X_A}{T_{2A}} + \frac{X_B}{T_{2B}} \tag{10.35}$$

where T_{2A} and T_{2B} denote the individual spin–spin relaxation times of protons in these two chemical environments.

The intermediate range, where chemical exchange takes place at a rate lying between the two extremes considered above, sees the NMR spectrum of this system change from one of two broadening lines to that of a single narrowing line at the frequency given in equation (10.34). The full expression for this has been evaluated but, for the general case, it is a complicated one. Let us make the following simplifying assumptions:

(i) Equal populations of protons in the two chemical environments, and thus equal lifetimes: i.e. $X_A = X_B = 0.5$ and $\tau_A = \tau_B = 2\tau$.
(ii) Large and equal spin–spin relaxation times, T_{2A} and T_{2B}, so that contributions to the line-width other than from chemical exchange may be ignored.

The formula for the line shape function, $g(\nu)$ then becomes,

$$g(\nu) = \frac{K\tau(\nu_A - \nu_B)^2}{\{\frac{1}{2}(\nu_A + \nu_B) - \nu\}^2 + 4\pi^2\tau^2(\nu_A - \nu)^2(\nu_B - \nu)^2} \tag{10.36}$$

where K is a constant.

The shape of this function depends only on the product, $\tau(\nu_A - \nu_B)$. The manner in which the shape changes from two separate lines centred at ν_A and ν_B when this product is large to one line at ν' when it is small is illustrated in Figure 10.5, which shows that the coalescence of the two peaks is achieved at $\tau(\nu_A - \nu_B) = (\sqrt{2}\pi)^{-1}$.

In using NMR spectroscopy to measure exchange rates, calculations of equation (10.36) at selected values of τ may be used to obtain the line shape showing optimum agreement with the actual spectrum. In this way it is possible to evaluate τ at each temperature at which the spectrum shows appreciable line-broadening due to exchange. Such τ values tend to lie in the range 1 to 10^{-3} s.

Kinetic studies by NMR are usually made on systems in chemical equilibrium. Thus the experimental data for τ refer, not to the depletion of the concentration of any species, but rather to the mean lifetime of individual molecules of that species. The total rate of reaction of this species, divided by its concentration, would give the pseudo-first order rate constant for its disappearance if the reverse steps were absent. The inverse of that quantity will represent the mean lifetime, τ.

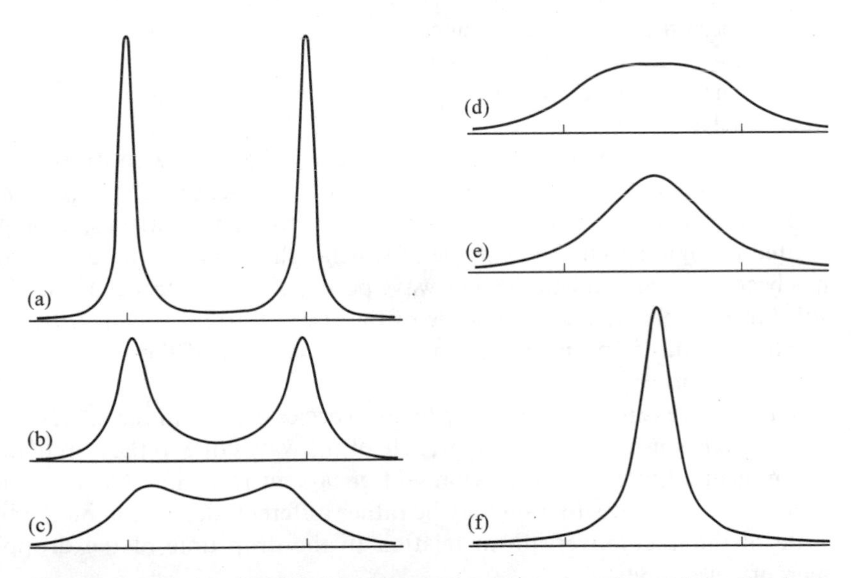

Fig. 10.5 Calculated NMR line shapes where proton exchange is involved, on the basis of equation (10.36), as a function of the ratio of τ^{-1} to $(\nu_A - \nu_B)$. (a) $\tau(\nu_A - \nu_B) = 6/\pi$; (b) $\tau(\nu_A - \nu_B) = 5/2\pi$; (c) $\tau(\nu_A - \nu_B) = 1/\pi$; (d) $\tau(\nu_A - \nu_B) = 1/\sqrt{2}\pi$; (e) $\tau(\nu_A - \nu_B) = 1/2\pi$; (f) $\tau(\nu_A - \nu_B) = 1/5\pi$.

In view of the similarities between NMR and electron spin resonance (ESR), line broadening in the latter technique is also indicative of exchange reactions. The major difference is that since ESR uses microwave radiation, of much higher frequency than the radio waves used in NMR, the accessible time scale is much shorter.

For example, the action of an alkali metal on naphthalene in tetrahydrofuran produces the naphthalenide radical anion, $C_{10}H_8^{\overline{\cdot}}$. If further naphthalene is added, then the exchange reaction,

$$C_{10}H_8^{\overline{\cdot}} + C_{10}H_8 \rightleftarrows C_{10}H_8 + C_{10}H_8^{\overline{\cdot}} \tag{10.37}$$

occurs and causes a broadening of the ESR lines. The rate constants so deduced are a function of the counter-ion involved, but lie in the region of 10^7 to $10^8\,dm^3\,mol^{-1}\,s^{-1}$. These do not represent the fastest reactions measurable by this technique, which can cope with rate constants several orders of magnitude higher.

10.5 Electrochemical methods

There are a number of electrochemical techniques applicable to the determination of fairly fast reactions in solution. The nature of these techniques imposes limitations on the scope of the reactions which they may be employed to study. Firstly, at least one reaction participant must take part in a redox process at the electrode; secondly, a solvent of fairly high relative permittivity is needed since it is necessary to work in a moderate concentration of an inert electrolyte.

Polarography, developed in the 1930s, is a widely used analytical technique. It solves the problem of electrode contamination by using a dropping mercury electrode, in which the surface is renewed every few seconds. As the negative potential of the mercury drop is slowly increased, the current normally rises in steps, each of which possesses two characteristics, as is illustrated in Figure 10.6. The voltage at which half of this increase in current has been attained is called the half-wave potential and is characteristic of the substance involved. The magnitude of the rise in current is called the wave height, and this is proportional to the concentration of the species undergoing reduction.

Where there is more than one reducible species present in the solution, the polarogram is usually just the aggregate of the waves due to these individual components. But if the species present engage in reversible reactions with each other, then the situation may be rather different, depending on the time scale of the interconversion in relation to the drop time of the dropping mercury electrode.

Suppose we have the isomerisation,

$$A \underset{-1}{\overset{1}{\rightleftarrows}} B \tag{10.38}$$

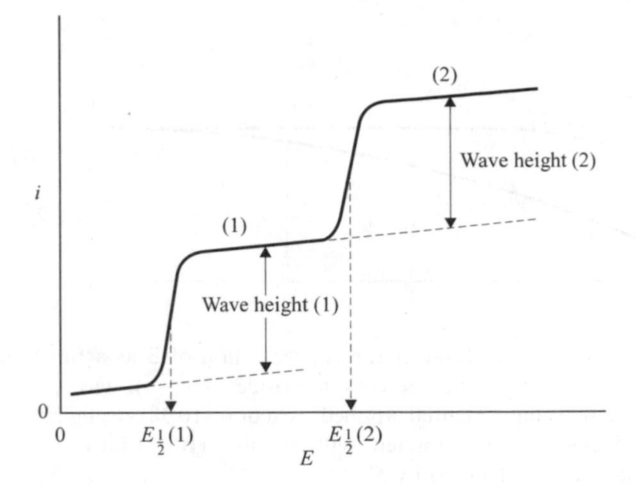

Fig. 10.6 A sketch of a polarogram, a plot of the current i against the (negative) electrode potential, E, showing two reduction waves and indicating their half-wave potentials, $E_{1/2}(1)$ and $E_{1/2}(2)$, and their wave heights.

where species B undergoes reduction at the dropping mercury electrode. It is not necessary that A be totally inactive at the cathode: it may well undergo reduction at a more negative potential than B. Let us assume that the equilibrium in reaction (10.38) lies towards the left, so that the concentration of B is very much less than that of A. When the dropping mercury electrode is held at a potential sufficiently negative to effect the reduction of species B, then the concentration of B will become depleted in this vicinity for the reason that every B molecule making contact with the cathode is effectively removed. At the cathode the concentration of B is essentially zero and around it there is a concentration gradient, reflecting the fact that the current flowing in the reduction wave is diffusion limited. The range of this depletion in concentration will be around the thickness, δ, of the Nernst diffusion layer. Where the rate constant k_1 is negligible, the forward step of reaction (10.38) will not cause any disturbance to this concentration gradient, so that the wave height realised will be a fair reflection of the concentration of B in the bulk of the solution.

If, on the other hand, the value of k_1 were moderately high, then the forward step of reaction (10.38) will cause the actual concentration of B in the vicinity of the cathode to be greater than that deduced from purely diffusional considerations, as is sketched in Figure 10.7. In consequence, the flux of species B towards the cathode is enhanced, so that the wave height is greater than would be anticipated on the basis of the mean concentration of B in the solution.

Thus, if reaction (10.38) cannot be neglected, the wave height of the reduction wave due to species B is a function, not merely of the concentration of B, but also of the concentration of A and of the rate constant k_1 for

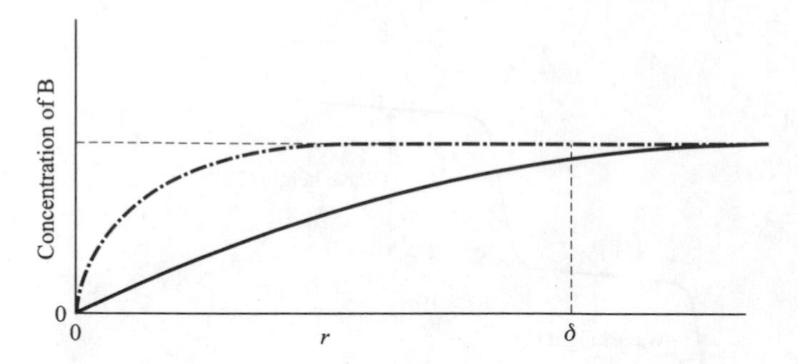

Fig. 10.7 Sketch of the variation of the concentration of B as a function of distance, r, from the dropping mercury electrode: (- - - -), zero potential applied; (——), reducing potential applied, reaction (10.38) negligibly slow; (– · – · –), reducing potential applied, with reaction (10.38) fairly fast. The Nernst diffusion layer is denoted by δ.

its conversion to B. Where the sum of the concentrations of A and B is fixed, it may be shown that the ratio of the mean cathode current i_k to the maximum mean cathode current, i_m, attained if the reaction were infinitely fast, is given by,

$$\frac{i_k}{i_m} = \frac{0.886(k_1 K_1 t)^{1/2}}{1 + 0.886(k_1 K_1 t)^{1/2}} \tag{10.39}$$

where K_1 denotes the equilibrium constant for reaction (10.38), t the drop time of the dropping mercury electrode and 0.886 is the constant factor which an advanced analysis has shown to be more accurate than the original formula of $(7/3\pi)^{1/2} = 0.862$. Thus, under conditions where the mean cathode current i_k is half the maximum value, we have:

$$k_1 = \frac{1}{(0.886)^2 K_1 t} = \frac{1}{0.785 K_1 t} \tag{10.40}$$

In many instances the chemical reaction involved is an acid–base process,

$$A^- + H^+ \underset{-a}{\overset{a}{\rightleftarrows}} HA \tag{10.41}$$

where the protonated species is the one readily undergoing reduction. The rate constant k_1 used in relation to reaction (10.38) now needs to be replaced by $k_a[H^+]$. In regard to reaction (10.41), it is more conventional to use the equilibrium constant for *dissociation* of HA, K_a, so that the ratio of the concentrations of HA to A^- will be given by $[H^+]/K_a$. Thus the corresponding equation in relation to reaction (10.41) is:

$$\frac{i_k}{i_m} = \frac{0.886[H^+](k_a t/K_a)^{1/2}}{1 + 0.886[H^+](k_a t/K_a)^{1/2}} \tag{10.42}$$

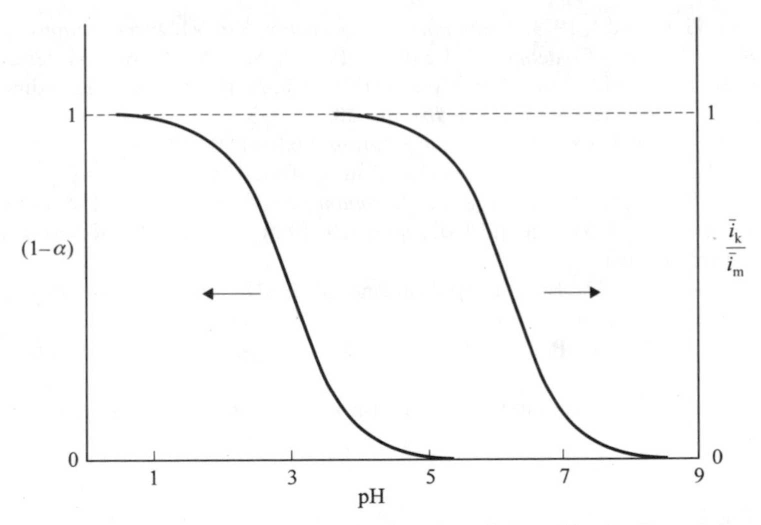

Fig. 10.8 The left hand curve is the fraction of the acid, of $pK_a = 3.0$, present as the neutral acid; the right hand curve is the ratio i_k/i_m for the same acid, taking $k_a = 10^9\,dm^3\,mol^{-1}\,s^{-1}$ and $t = 3\,s$, both plotted against pH.

So, when i_k is half of i_m, k_a will be given by:

$$k_a = \frac{K_a}{0.785\,t\,[H^+]^2} \tag{10.43}$$

The variation of equation (10.42) with pH is shown in Figure 10.8, taking K_a as 10^{-3} mol dm^{-3}, k_a as $1 \times 10^9\,dm^3\,mol^{-1}\,s^{-1}$ and t as 3 s. This shows that whereas i_k is almost equal to i_m at pH 4, it decreases with increasing pH, attaining 0.5 at pH 6.19 and going virtually to zero at pH 8. On the same graph there is plotted the fraction of the acid present as the protonated species, which obviously crosses 0.5 at pH 3.0. The discrepancy between these two curves is the consequence of reaction (10.41) being much too fast to be neglected.

Polarography would seem an obvious technique to use in measuring the rates of protonation of carboxylates since, almost without exception, these acids are more readily reduced than are the corresponding anions. However, the values thus obtained are susceptible to various sources of error. For pyruvic (2-oxo-propanoic) acid, whose pK_a is 2.40, the value for k_1 was found to be $1.3 \times 10^{10}\,dm^3\,mol^{-1}\,s^{-1}$.

There are a great many other electrochemical techniques and variations of techniques which may be applied to kinetic problems, but it is not possible to do them justice within the scope of this book.

Suggested reading

Bernasconi, C. F., 1976. *Relaxation Kinetics*. New York: Academic Press.

Hammes, G. G. (ed.), 1974. *Techniques of Chemistry*, Vol VI. *Investigations of rates and mechanisms of reactions*, 3rd edition, Part 2. New York: Wiley-Interscience.
Bannister, J. J., J. Gormally, J. F. Holzwarth and T. A. King, 1984. The iodine laser and fast reactions. *Chemistry in Britain*, **20**, 227–33.
Caldin, E. F., 1964. *Fast Reactions in Solution*. Oxford: Blackwell.
Bradley, J. N., 1975. *Fast Reactions*. London: Oxford University Press.
Harris, R. K., 1986. *Nuclear Magnetic Resonance Spectroscopy*. London: Longman.
Martin, M. L., G. J. Martin and J.-J. Delpuech, 1980. *Practical NMR Spectroscopy*. London: Heyden.
Lincoln, S. F., 1977. Kinetic applications of NMR spectroscopy. *Progress in Reaction Kinetics*, **9**, 1–91.
Zuman, P., and R. C. Patel, 1984. *Techniques in Organic Reaction Kinetics*. New York: Wiley.
Zuman, P., 1967. Polarography and reaction kinetics. *Advances in Physical Organic Chemistry*, **5**, 1–52.

Problems

10.1 For the reaction system,

$$A + A \underset{-1}{\overset{1}{\rightleftarrows}} A_2$$

deduce the expression for the relaxation time (assuming a small perturbation from equilibrium) in terms of the rate constants for the forward and reverse processes and the equilibrium concentration of A.

10.2 For the reaction of 2,4-dinitrophenol (AH) with tri-1-octylamine (B),

$$AH + B \underset{-1}{\overset{1}{\rightleftarrows}} A^- \ldots HB^+$$

to form an ion-pair, the equilibrium constant in chlorobenzene at 297 K is known to be *ca* 4.8×10^3. For solutions with the following *nominal* concentrations of acid and base, the relaxation times at 297 K were found to be as listed.

$[AH]_0/10^{-5}$ mol dm^{-3}	$[B]_0 10^{-5}$ mol dm^{-3}	$\tau/\mu s$
9.6	13.3	10.8
12.8	20.0	10.2
12.8	26.6	9.0
12.8	33.3	8.2
16.0	40.0	7.0
19.2	46.6	6.65
25.65	60.0	5.8

Determine the values of k_1 and k_{-1} for the above reaction. (Data from Caldin, E. F., J. E. Crooks and D. O'Donnell, *J. Chem. Soc., Faraday Trans I*, 1973, **69**, 993–9.)

10.3 For the reaction scheme,

$$A + B \underset{-1}{\overset{1}{\rightleftarrows}} C \underset{-2}{\overset{2}{\rightleftarrows}} D$$

a possible circumstance is that reaction steps 2 and –2 are so fast that species C and D are in effective equilibrium. Deduce the expression for the relaxation time in terms of the rate constants k_1 and k_{-1}, the equilibrium constant K_2 and the equilibrium concentrations of A and B.

10.4 When the phenol tetrabromophenolphthalein ethyl ester is added to a substituted pyridine in chlorobenzene solution, a new absorption band appears at 560 nm, attributable to the product of the proton transfer reaction. Careful measurements showed that the system does not fit the simple scheme

$$AH + B \rightleftarrows D$$

but is consistent with the scheme,

$$AH + B \overset{1}{\rightleftarrows} C \overset{2}{\rightleftarrows} D$$

where C represents a hydrogen bonded complex between the phenol and the base.

Kinetic studies of this system using a microwave temperature jump apparatus showed only a single relaxation time, and the reciprocal of τ varied linearly with the sum of the equilibrium concentrations of AH and B. Assuming that either step 1 or step 2 is too fast to be detected in these experiments, discuss the possible significance to be attributed to the slope and intercept of these plots.

10.5 Polarographic studies of pyruvic acid in slightly acid solutions show a reduction wave due to undissociated pyruvic acid and one due to pyruvate anion, in proportions which are pH-dependent. With a drop time of 3 seconds, the wave due to pyruvate anion was reported to fall to half of its maximum value in a buffered solution at pH 6.12, whereas the pK_a of pyruvic acid is 2.50. Thus evaluate the rate constant for the reaction of H^+ + pyruvate. (Data from Delahay, P., *New Instrumental Methods in Electrochemistry*, 1954, New York, Interscience.)

10.6 For the system,

$$A \underset{-1}{\overset{1}{\rightleftarrows}} B \underset{-2}{\overset{2}{\rightleftarrows}} C$$

in which step 1 is much more rapid than step 2, derive the two relaxation times in terms of the relevant rate constants and the equilibrium concentrations of A, B and C.

11 Photochemistry and radiation chemistry

Subjecting a substance or a mixture of substances to visible or UV light may induce a chemical reaction, or alter the rate of a reaction already occurring. This topic, called *photochemistry*, is an extensive one, with many facets in addition to the kinetic considerations of relevance here.

The existence of various ionising radiations was first recognised in the 1890s, with the discovery of X-rays and then of emissions from radioactive decay, in both cases detected by the chemical effect of the fogging of a photographic plate. Since then, there have been extensive studies of the chemical effects of ionising radiation, a topic known as *radiation chemistry*.

The common theme of photochemistry and radiation chemistry is that both generate unstable species and thus offer means of observing their reactions. Some of these species are so reactive that this must occur on a very short time scale, so appropriate fast reaction techniques have been devised. By providing a route to study the behaviour and the reactivity of species not usually present, these topics have significantly enhanced our knowledge of chemistry.

11.1 Initial effects of light absorption

One of the earliest significant observations regarding the chemical aspects of light was that of Grotthus and Draper that "only light which is absorbed can produce photochemical change". However, the absorption of light in the UV or visible can have any one of a large number of possible consequences.

In most cases, molecules have no unpaired electrons and thus have a singlet ground state S_0. So the electronic transition caused by light absorption by the molecule, AB, produces an excited singlet state of this molecule, AB(S_n), for which some of the possible fates are illustrated in Figure 11.1.

Step (i), covalent bond dissociation, may ensue from the elevation of an electron from a bonding or a non-bonding orbital to an anti-bonding orbital, resulting in a state which is either weakly bound or repulsive, so that

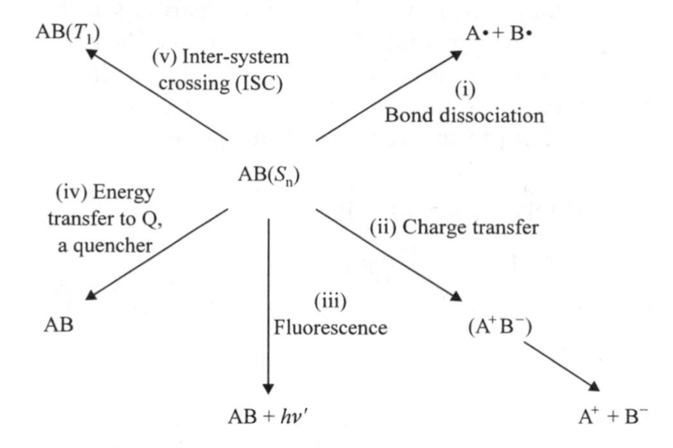

Fig. 11.1 Some possible photochemical and photophysical consequences for the molecule AB, arising from its promotion to an excited singlet state by the absorption of a quantum of light.

homolytic bond rupture occurs within one vibrational period. The immediate consequences can thus be represented as, for example:

$$I_2 \xrightarrow{h\nu} I^{\cdot} + I^{\cdot} \tag{11.1}$$

Step (ii) is readily illustrated by an ion pair possessing an electronic transition of a charge-transfer character, such as:

$$Fe^{3+}Cl^{-} \xrightarrow{h\nu} (Fe^{2+}Cl^{\cdot}) \rightarrow Fe^{2+} + Cl^{\cdot} \tag{11.2}$$

As a consequence of light absorption, a ferrous ion and a chlorine atom are produced in close proximity. Since these do not mutually attract, they may diffuse apart once the solvent cage permits and thereafter behave independently.

There are many examples of charge-transfer processes among the coordination compounds of transition metals. Whether these are metal-to-ligand or ligand-to-metal transitions, typically they result in a different oxidation state of the metal and an unstable species derived by the reduction or the oxidation of the ligand.

In the absence of channels of dissociation, the original excited singlet state will usually undergo *internal conversion* to the lowest excited singlet state, S_1. This may then return to the ground state by the emission of a quantum of light, $h\nu'$, a process known as *fluorescence*, step (iii). Alternatively, the excess energy may be passed to another molecule, Q, which does not fluoresce: this is known as fluorescence quenching, step (iv).

Where there is a triplet state of the molecule AB at an energy close to that of the lowest excited singlet state, S_1, this may be formed in a process known as *inter-system crossing* (ISC) (step (v)). The lowest triplet state, T_1, is in general much longer-lived than an excited singlet state, but it may, on account of its electron configuration and electronic distribution, be much more reactive than the singlet ground state. For example, benzophenone dissolved in a 2-propanol is quite unreactive. Under UV excitation, the triplet state of benzophenone is formed, and it may abstract a hydrogen atom from 2-propanol, thus producing two radical species:

$$\left.\begin{array}{rcl} \mathrm{PhCOPh}(S_o) & \stackrel{h\nu}{\rightarrow} & \mathrm{PhCOPh}(S_1) \\ \mathrm{PhCOPh}(S_1) & \stackrel{\mathrm{ISC}}{\rightarrow} & \mathrm{PhCOPh}(T_1) \\ \mathrm{PhCOPh}(T_1) + \mathrm{Me_2CHOH} & \rightarrow & \mathrm{Ph_2\dot{C}OH} + \mathrm{Me_2\dot{C}HOH} \end{array}\right\} \tag{11.3}$$

Reference to monographs on photochemistry listed at the end of the chapter will permit this list of the processes which might follow light absorption to be augmented and refined. However, in regard only to the *kinetics* of photochemical reactions, it may well be claimed that Figure 11.1 constitutes an adequate summary.

The efficiency of absorbed light in bringing about chemical change is usually reported in terms of a ratio, defined as the *quantum yield,* and denoted by ϕ. We have

$$\begin{aligned} \phi &= \frac{\text{number of product molecules formed}}{\text{number of light quanta absorbed}} \\ &= \frac{\text{number of moles of product formed}}{\text{number of Einsteins of light absorbed}} \end{aligned} \tag{11.4}$$

where the second line is obtained by multiplying numerator and denominator by Avogadro's constant.

A more recent photochemical "law", that of Einstein and Stark, states that precisely one molecule is excited for each quantum absorbed. It does not then follow, even for monophotonic processes, that the quantum yield of every photochemical reaction is necessarily unity. For this there are several reasons, including the following:

(i) If the primary process is homolytic bond rupture of a species in solution, then some fraction of the pairs of radicals thus generated may be expected to recombine within the solvent cage, i.e. to undergo geminate recombination.

(ii) If, in the gas phase, one I_2 molecule is dissociated for every quantum absorbed then the quantum yield of I atoms is equal to two. For such a reaction we must be clear whether by quantum yield we mean $\phi(\mathrm{I})$ or $\phi(-\mathrm{I_2})$, since these quantities differ by a factor of two.

(iii) If a species generated photochemically should become the chain carrier in a chain reaction, then the quantum yield of the eventual product could be several powers of ten.

11.2 Photochemical kinetics

In this section, various aspects of photochemical studies are surveyed to demonstrate the kinetic information obtainable and the type of problem which may have to be surmounted.

11.2.1 Fluorescence quenching

The photophysical process of the fluorescence of an excited singlet state has been referred to in the previous section. In general, it will take place in parallel with non-radiative decay processes, also leading to the ground state. Denoting the polyatomic molecule by A and its excited singlet state by A^*, these events may be summarised as follows:

$$\left.\begin{array}{lcl} A & \xrightarrow{h\nu} & A^* \\ A^* & \xrightarrow{f} & A + h\nu' \\ A^* & \xrightarrow{g} & A \end{array}\right\} \qquad (11.5)$$

The addition of certain substances causes a decrease in the fluorescence intensity, even though the substance is transparent at the wavelength of excitation. This may happen because of energy transfer from the excited singlet to the other molecule, Q, which then sheds its excess energy in a radiationless process:

$$\left.\begin{array}{rcl} A^* + Q & \xrightarrow{q} & A + Q^\dagger \\ Q^\dagger & \rightarrow & Q \end{array}\right\} \qquad (11.6)$$

The rate of the excitation process is given by the product of the quantum yield ϕ of forming the excited state A^*, and I_a, the rate of light absorption, in Einsteins per unit volume per unit time, which is a function of the concentration of A. Thus we have, for the net rate of formation of A*,

$$\frac{d[A^*]}{dt} = \phi I_a - k_f[A^*] - k_g[A^*] - k_q[A^*][Q] \qquad (11.7)$$

This gives, for the steady state concentration:

$$[A^*] = \frac{\phi I_a}{k_f + k_g + k_q[Q]} \qquad (11.8)$$

Since the intensity of the fluorescence is given by k_f[A*] and so is proportional to the concentration of the excited singlet, the ratio of the fluorescence intensity I_f^o in the absence of the quencher to the value I_f in its presence is given by:

$$\begin{aligned}\frac{I_f^o}{I_f} &= \frac{\phi I_a/(k_f + k_g)}{\phi I_a/(k_f + k_g + k_q[Q])} \\ &= 1 + \frac{k_q}{k_f + k_g}[Q]\end{aligned} \tag{11.9}$$

This is known as the Stern–Volmer equation. In the absence of a quencher, the mean lifetime τ_f of the excited singlet, the fluorescence lifetime, is given by $(k_f + k_g)^{-1}$. Thus the equation is sometimes written as,

$$\frac{I_f^o}{I_f} = 1 + k_q\tau_f[Q] \tag{11.9a}$$

and it implies that a plot of I_f^o/I_f against the concentration of quencher should be linear, with an intercept of one and the slope equal to $k_q\tau_f$. If τ_f can be measured in separate experiments then the value of k_q can be obtained. Frequently, it lies close to the diffusion-controlled limit. A typical Stern–Volmer plot is shown in Figure 11.2.

The energy transfer mechanism described above is found not to be operative in all cases of fluorescence quenching. Where the pair of compounds involved constitute an electron donor and an electron acceptor then, especially in a medium of moderate dielectric permittivity, the process involved may well be charge transfer:

$$A^* + Q \quad \rightarrow \quad A^{(\mp)} \ldots Q^{(\pm)} \tag{11.10}$$

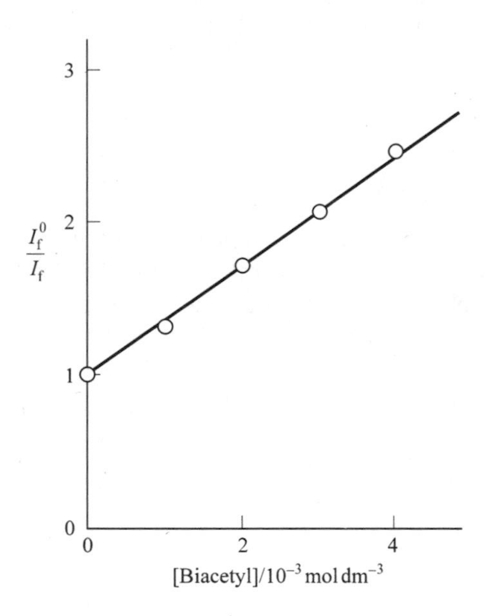

Fig. 11.2 A Stern–Volmer plot for the fluorescence quenching of chrysene by biacetyl in cyclohexane solution, where I_f^o/I_f is plotted against the concentration of quencher. The straight line gives $k_q\tau_f = 3.6 \times 10^2\, dm^3\, mol^{-1}$. (Data from Stevens, B. and J. T. Dubois, *Trans Faraday Soc.*, 1963, **59,** 2813–19.)

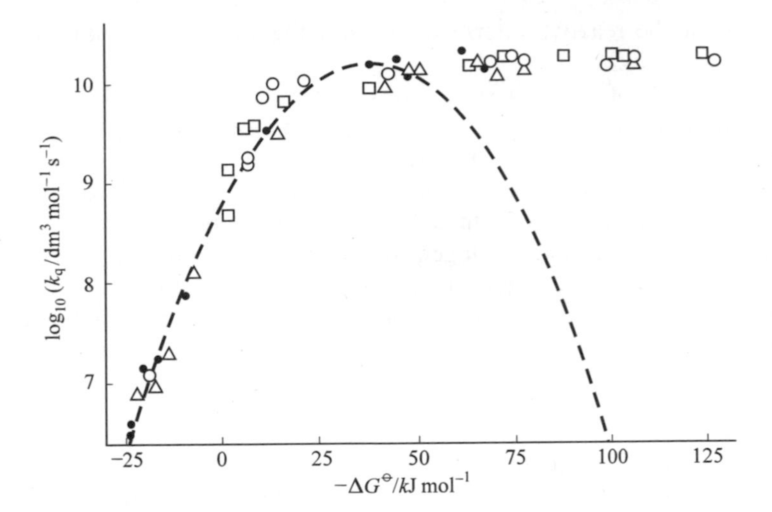

Fig. 11.3 Plot of $\log_{10} k_q$ against $-\Delta G^{\ominus}$ for a group of reactions in acetonitrile effecting fluorescence quenching by electron transfer. Fluorescing compounds: pyrene-3-carboxylic acid, (○); benz(c)acridine, (□); benz(a)anthracene, (Δ); pyrene-4-carboxylic acid, (●). The dashed line is calculated from equation (5.37a) (the Marcus equation). (Data from Rehm, D. and A. Weller, *Israel J. Chem.*, 1970, **8**, 259–71.)

Data on quenching reactions of this type, occurring in acetonitrile, were presented in 1970 by Rehm and Weller, where the fluorescing compound was an electron acceptor and the quencher an electron donor. By measuring the half-wave potentials for reduction of the fluorescing compound and for oxidation of the quencher, along with the excitation energy of the fluorescing singlet state, the standard Gibbs energy change of each quenching reaction, $\Delta G_q^{\ominus}$, was evaluated.

The quenching rate constants are plotted, on a log scale, in Figure 11.3 as a function of $\Delta G_q^{\ominus}$. This plot demonstrates that the predictions of Marcus theory (see Section 5.7) are obeyed by these electron transfer reactions when $\Delta G_q^{\ominus}$ is positive or slightly negative. However, for more negative values of $\Delta G_q^{\ominus}$, the rate of the quenching reaction remains close to the diffusion-controlled limit and does not decline in the Marcus inverted region as the theory would predict.

11.2.2 Competition studies

In a photochemical system, it is sometimes possible to generate a reactive intermediate capable of undergoing reaction with either of two solutes, to yield different respective products. By measuring how the yields of the products vary with the concentrations of the two solutes, the relative rate constant of the reactive intermediate with the two solutes may be determined (*cf.* p. 50).

One example of such a system utilises a halide ion, Br^- or I^-, in aqueous solution. Halide ions absorb in the UV in a manner which shows considerable dependence on the solvent and the transition is attributed to a charge-transfer-to-solvent absorption, where an electron is excited to an orbital within the potential well formed by the oriented polar molecules of the solvent. Separation of the halogen atom and the electron can then occur, with an efficiency which depends on the solvent and which rises with temperature, but is typically in the range 0.2 to 0.4. Thus the process can be represented by,

$$X^- \xrightarrow{h\nu} X^{\cdot} + e^-_{solv} \tag{11.11}$$

with the appropriate quantum yield.

It is convenient to use N_2O as one of the solutes with which the electron may react. It is transparent at the wavelengths used to excite I^- or Br^- ions and, on reaction, it results in the formation of a readily measurable product. In aqueous iodide the reactions are:

$$\left.\begin{array}{rcl} I^- & \xrightarrow{h\nu} & I^{\cdot} + e^-_{aq} \\ e^-_{aq} + N_2O & \xrightarrow{1} & N_2 + O^- (\xrightarrow{H^+} OH) \\ OH + I^- & \rightarrow & OH^- + I^{\cdot} \\ I^{\cdot} + I^- & \rightarrow & I_2^- \\ I_2^- + I_2^- & \rightarrow & I_3^- + I^- \end{array}\right\} \tag{11.12}$$

Under these circumstances, $\phi(N_2) = \phi(I_3^-) = \phi_{e^-}$. If another transparent solute, such as an alkyl chloride, is added, then we have the competing reaction:

$$e^-_{aq} + RCl \xrightarrow{2} Cl^- + R^{\cdot} \tag{11.12a}$$

Consequently, the quantum yield of N_2 is decreased. The equation may be obtained formally by considering the steady state concentration of the hydrated electron, e^-_{aq}:

$$\phi(N_2) = \phi_{e^-}\left\{\frac{k_1[N_2O]}{k_1[N_2O] + k_2[RCl]}\right\} \tag{11.13}$$

Inverting this equation, we obtain:

$$\frac{1}{\phi(N_2)} = \frac{1}{\phi_{e^-}}\left\{1 + \frac{k_2[RCl]}{k_1[N_2O]}\right\} \tag{11.14}$$

Thus, $\{\phi(N_2)\}^{-1}$ should be a linear function of the concentration ratio $[RCl]/[N_2O]$, when both are varied. The relative rate constant, k_2/k_1, is

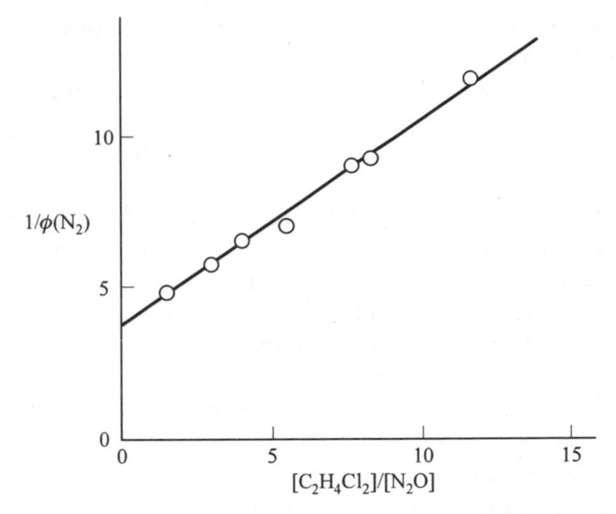

Fig. 11.4 Plot of the reciprocal of $\phi(N_2)$ against the concentration ratio [1,2-dichloroethane]/[N_2O], from studies in which the iodide ion was excited by light of 253.7 nm.

determined as the ratio of the slope to the intercept, as is shown in Figure 11.4.

In using equation (11.14), it is important to measure not merely the yield of N_2 over a certain period of illumination, but the actual quantum yield. This requires that the rate of light input to the cell is measured, either just before or just after the irradiation of the sample containing iodide ion, and this may be achieved using chemical actinometry.

The cell used for actinometry should be of the same size, shape and material as that used for the illumination of the sample. It is usual to operate with total absorption of the light entering the cell. To achieve this, it is necessary that, at the excitation wavelength, the absorbance $A = \varepsilon cl$ is at least 3, corresponding to the absorption of 99.9% of the incident light.

One of the mostly widely used actinometers is the ferrioxalate system, containing the ion $[Fe^{3+}(C_2O_4{}^{2-})_3]^{3-}$ in dilute sulphuric acid solution. A ferrioxalate concentration of 0.006 mol dm^{-3} is sufficient to give complete absorption with a 1 cm path length, over a range of wavelengths in the ultra-violet.

When the ferrioxalate ion absorbs a quantum of light, the process occurring is a ligand-to-metal charge transfer, which may be represented as:

$$\left.\begin{aligned} [Fe^{3+}(C_2O_4^{2-})_3]^{3-} &\xrightarrow{h\nu} [Fe^{2+}(C_2O_4^{2-})_2(C_2O_4^{-})]^{3-} \\ [Fe^{2+}(C_2O_4^{2-})_2(C_2O)_4^{-})]^{3-} &\rightarrow Fe^{2+} + 2C_2O_4^{2-} + CO_2 + CO_2^{\dot{-}} \\ CO_2^{\dot{-}} + [Fe^{3+}(C_2O_4^{2-})_3]^{3-} &\rightarrow CO_2 + Fe^{2+} + 3C_2O_4^{2-} \end{aligned}\right\} \quad (11.15)$$

Thus as a consequence, some ferric ions are reduced to ferrous, with the concomitant production of CO_2.

It is quite easy to estimate Fe^{2+} in the presence of Fe^{3+}, using the fact that the former forms a red complex with 1,10-phenanthroline, with an absorption peak in the visible at 510 nm, where there is negligible absorption from the ferrioxalate ion.

The sensitivity of the ferrioxalate system owes much to the fact that one directly measures the amount of ferrous ion produced. By contrast, in the uranyl oxalate system one monitors reaction only by measuring the relatively small decrease in the total amount of oxalate present.

The quantum yield of the ferrioxalate actinometer, as determined by reference to primary standards, is found to show some variation with wavelength, but over the range 250–440 nm, it lies between 1.26 and 1.10. The fact that the quantum yield is not equal to 2.0 reflects the occurrence of back-reactions within the cage, so that not every quantum absorbed by ferrioxalate ion generates one ferrous ion and a $CO_2^{\cdot -}$ radical ion able to produce another.

11.2.3 Gas phase kinetics

The use of photochemical processes in the study of gas phase kinetics is exemplified by the photolysis of gaseous azomethane in a closed vessel at slightly elevated temperatures, using filtered light of wavelengths in excess of 300 nm. The major products are nitrogen, methane and ethane and it would seem likely that the following steps occur:

$$\left.\begin{aligned} CH_3N_2CH_3 &\xrightarrow{h\nu} CH_3N_2^{\cdot} + CH_3^{\cdot} \\ CH_3N_2^{\cdot} &\xrightarrow{1} CH_3^{\cdot} + N_2 \\ CH_3^{\cdot} + CH_3N_2CH_3 &\xrightarrow{2} CH_4 + \dot{C}H_2N_2CH_3 \\ CH_3^{\cdot} + CH_3^{\cdot} &\xrightarrow{3} C_2H_6 \end{aligned}\right\} \tag{11.16}$$

On this basis, the rates of formation of methane and ethane can be expressed as follows:

$$R_{CH_4} = k_2[CH_3^{\cdot}][CH_3N_2CH_3] \tag{11.17}$$

$$R_{C_2H_6} = k_3[CH_3^{\cdot}]^2 \tag{11.18}$$

By taking the square root of the second equation, and then dividing equation (11.17) by it, the methyl radical concentration may be eliminated, leading to:

$$\frac{R_{CH_4}}{(R_{C_2H_6})^{1/2}} = \frac{k_2[CH_3N_2CH_3]}{k_3^{1/2}} \tag{11.19}$$

Thus, from the experimental values of the rates of formation (in moles per unit volume per unit time) of these alkane products and the known concentration of azomethane, it is possible to evaluate the ratio, $k_2/k_3^{1/2}$. Combining measurements over a range of temperatures with an absolute value for k_3, determined from other work, leads to the Arrhenius parameters of k_2. This system can also lend itself to the study of reactions of the methyl radical with other molecules, provided that they are transparent to the wavelengths of light used to photodissociate the azomethane.

11.2.4 Spatial inhomogeneity effects

When a photochemical process is used to initiate a chemical reaction with a view to studying its kinetics, one fact of potential importance is that, almost inevitably, the light absorption does not occur uniformly so that the reactive intermediates are not produced uniformly throughout the volume of the cell. This fact is especially important where there is a step in the reaction process in which one reaction intermediate reacts with another. Also, it is a much more important consideration in solution than in the gas phase, where diffusion and convection will be more effective in levelling out the concentrations.

In this treatment, it will be assumed that the photolysing light is monochromatic and that the decadic absorption coefficient of the photolyte is ε at the wavelength of illumination, whereas the solvent and any other solutes present do not absorb. Let us suppose that a beam of light of Q Einstein cm^{-2} s^{-1} impinges at normal incidence on the front face of the cell. If l is the path length of the cell, at a distance $x \leq l$ from the front face of the cell, the rate of light absorption, I_a, will be given by,

$$\begin{aligned} I_a &= -\frac{d}{dx}(Q.10^{-\varepsilon cx}) \text{ Einstein cm}^{-3}\,\text{s}^{-1} \\ &= 2.303 \times 10^3\,\varepsilon c Q.10^{-\varepsilon cx} \text{ Einstein dm}^{-3}\,\text{s}^{-1} \end{aligned} \tag{11.20}$$

where c denotes the molar concentration of the photolyte. Clearly if the absorbance $A = \varepsilon cl$ is much less than one, the rate of light absorption is low and not so dependent on the value of x, whereas if $A \gg 1$, the rate of light absorption starts high and falls virtually to zero, as is illustrated in Figure 11.5.

The type of reaction in which the inhomogeneity of the light absorption is especially important is one where the rate of reaction, locally, depends on the rate of the initial photoreaction to a power other than one. This arises, for example, where a free radical chain polymerisation reaction is initiated photochemically.

Suppose an initiator, I, is converted photochemically into two radicals, R$^{\cdot}$, which react with the olefin, CH_2:CHX,

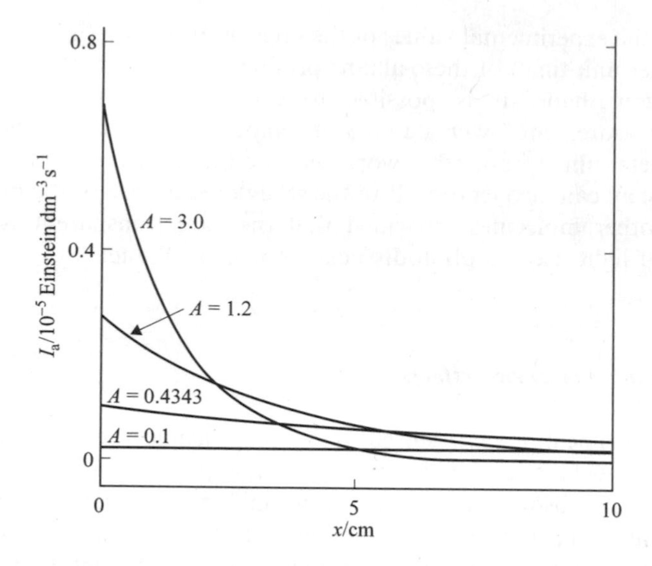

Fig. 11.5 Plots of I_a against x for the indicated values of $A = \varepsilon cl$, illustrating how I_a varies with concentration and with position in the cell. The ordinate values relate to $Q = 10^{-8}$ Einstein cm^{-2} s^{-1} and $l = 10$ cm. These curves illustrate that the highest I_a value at the rear of the cell is obtained with $A = 0.4343$.

$$\left.\begin{array}{rcl} \mathrm{I} & \xrightarrow{h\nu} & \mathrm{R^{\cdot} + R^{\cdot}} \\ \mathrm{R^{\cdot} + CH_2{:}CHX} & \xrightarrow{\mathrm{b}} & \mathrm{R.CH_2.\dot{C}HX}(m_1^{\cdot}) \end{array}\right\} \tag{11.21}$$

after which the reaction proceeds as outlined on schemes (7.30) and (7.31), where thermal generation of $R^{\cdot}$ was assumed. Consequently we obtain, instead of equation (7.35), the relation:

$$-\frac{\mathrm{d[CH_2{:}CHX]}}{\mathrm{d}t} = k_\mathrm{p}[m_x^{\cdot}][\mathrm{CH_2{:}CHX}]$$

$$= k_\mathrm{p}\left(\frac{\phi I_\mathrm{a}}{2k_\mathrm{t}}\right)^{1/2}[\mathrm{CH_2{:}CHX}] \tag{11.22}$$

Here ϕ denotes the quantum yield of the radical $R^{\cdot}$ which participates in reaction step b.

Whereas for thermal initiation, the rate of generating $m_1^{\cdot}$ is uniform throughout the cell, for photo-initiation it varies because of the non-uniformity of I_a and the absence of any rapid and efficient stirring to maintain a uniform concentration of $m_x^{\cdot}$ throughout the cell. To obtain an expression for the rate of consumption of monomer molecules within the cell, which is effectively what is monitored using a technique such as dilatometry, we need to sum over the whole cell. This gives, for the number of moles of monomer consumed within unit time,

$$R_M = \int_0^l k_p \left(\frac{\phi I_a}{2k_t}\right)^{1/2} [CH_2{:}CHX]\, 10^{-3}\, dx \qquad (11.23)$$

where the factor of 10^{-3} reflects the conversion from cm^3 to dm^3. Substituting for I_a from equation (11.20) and integrating, we obtain:

$$R_M = k_p CH_2{:}CHX] \left(\frac{\phi Q}{2k_t}\right)^{1/2} \frac{2}{(2.303 \times 10^3 \varepsilon c)^{1/2}} (1 - 10^{-\varepsilon c l/2}) \qquad (11.24)$$

Let us consider how this equation behaves in each of the two extreme regions of the absorbance, A. Where this is low so that $\varepsilon cl \ll 1$, the last factor in equation (11.24) corresponds, to a good approximation, to 2.303 $\varepsilon cl/2$, which gives:

$$R_M = k_p[CH_2{:}CHX] \left(\frac{2.303 \varepsilon c \phi Q}{2 \times 10^3 k_t}\right)^{1/2} l \qquad (11.24a)$$

So under these conditions, R_M is proportional to $c^{1/2}$.

If, on the other hand, the absorbance is high at the wavelength of irradiation so that $\varepsilon cl \gg 1$, then the last factor in equation (11.24) is effectively equal to one, which gives:

$$R_M = k_p[CH_2{:}CHX] \left(\frac{2\phi Q}{2.303 \times 10^3 \varepsilon c k_t}\right)^{1/2} \qquad (11.24b)$$

Thus under these conditions R_M is proportional to $c^{-1/2}$.

So, if the above reaction is carried out with Q and the monomer concentration kept constant, a plot of log R_M against log c will have two linear regions, with a slope of 0.5 at lower values of c and of -0.5 at higher c. The position of the intervening maximum, located by differentiating equation (11.24) with respect to c, is given by:

$$\ln(1 + 2.303 \varepsilon cl) = 2.303\, \varepsilon cl/2 \qquad (11.25)$$

This equation is satisfied when $\varepsilon cl = 1.0913$.

It is possible to offer a qualitative explanation for the behaviour at high concentrations of the initiator. As Figure 11.5 demonstrates, where the absorbance A is high, nearly all the light is absorbed in the first few millimetres of the cell, which is where the bulk of the reaction must occur. Thus the effective volume for the reaction is inversely proportional to A, and consequently proportional to c^{-1}. Within this effective volume, the mean value of I_a rises in proportion to c, but as equation (11.22) shows, the mean rate is proportional to $c^{1/2}$. The product of this factor of $c^{1/2}$ in the rate and c^{-1} in the volume yields the overall factor of $c^{-1/2}$ deduced above.

In the literature, there has sometimes been an unfortunate confusion between the rate of light input and the rate of light absorption. The former is what is measured in an actinometry experiment, and it may reasonably be held constant in a series of experiments. The latter is rarely constant within the cell, but if it is to remain unchanged in a series of experiments then it is paramount that the initial and the final concentrations of the light-absorbing species remain unaltered.

11.3 Flash photolysis

Where the absorption of light causes the occurrence of a chemical reaction, one must expect at least one reactive intermediate to be produced by the action of light, with perhaps others formed in subsequent reaction steps. Under continuous illumination, the concentrations of all these species would normally remain too low to permit their detection. The initial conception of flash photolysis was as a means of putting a large amount of light into a sample in a very short time, thus producing a sufficiently large concentration of each intermediate that these species may be monitored over a short time scale. The technique, pioneered in the late 1940s by Norrish and Porter, who were recognised in the award of the 1967 Nobel Prize, has been widely used in kinetic studies.

In the conventional design of a flash photolysis system, the flash is produced by the controlled discharge of a capacitor through two flash tubes, containing an inert gas such as krypton or xenon at a low pressure. The flash tubes are positioned parallel to the cell and are enclosed in a cylindrical reflector, so as to maximise the fraction of the emitted light entering the cell, as is illustrated in Figure 11.6. For kinetic experiments in solution, the usual

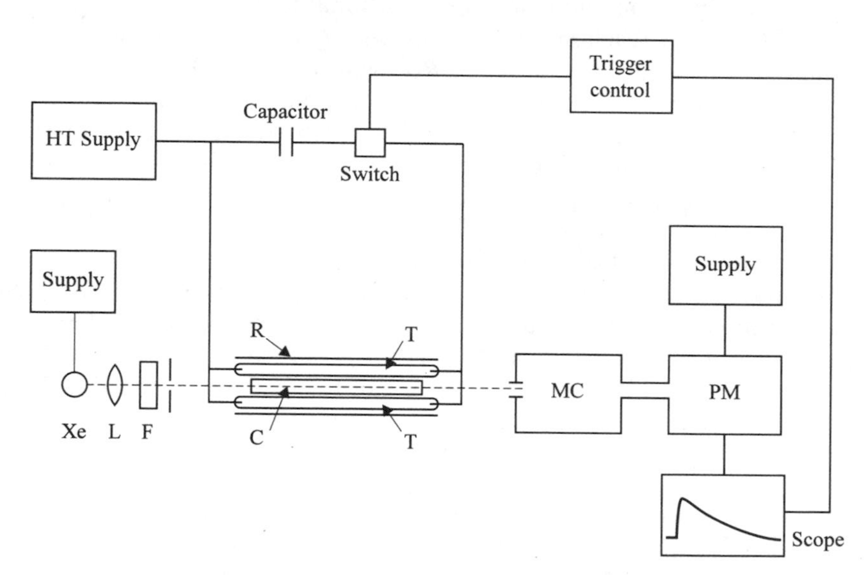

Fig. 11.6 Schematic diagram of a conventional flash photolysis apparatus, using spectrophotometric detection Xe, xenon lamp; L, lens; F, filter cell; C, reaction cell; T, flash tubes; R, reflector; MC, monochromator: PM, photomultiplier. An ignitron may be used as the switch for the capacitor discharge.

mode of detection is by kinetic spectrophotometry. A beam of white light from a xenon arc lamp, with a filter interposed to prevent its causing photolysis of the sample, is directed along the axis of the cell to a monochromator, and the light at the selected wavelength is fed to a fast response photomultiplier tube. Then using an oscilloscope triggered at the commencement of the flash, the changing amount of light transmitted through the cell is recorded as an oscilloscope trace.

The considerations involved in choosing the concentration of the species being photolysed are implicit in Figure 11.5. It is desired to have a high rate of light absorption at the axis of the cell, along which is directed the analysing beam. At high concentrations, I_a is very high at the edge of the cell so that scarcely any light penetrates to the axis. At very low concentrations, a high proportion of the light will reach that far but scarcely any of it will be absorbed there. By differentiating equation (11.20) with respect to c, it can easily be shown that the optimum condition is that $2.303\varepsilon cr = 1$, where r denotes the radius of the cell, or that $c = 0.4343/\varepsilon r$. The application of this formula is complicated by the fact that the light from the flash is not monochromatic, so that the chemical effects are caused by light over a range of wavelengths. Thus ε becomes a notional range rather than a precise value. Also, not all the effective light enters the cell at normal incidence. Consequently, the effective value of r is rather greater than the actual dimension of the cell.

Kinetic applications of a conventional flash photolysis apparatus may be illustrated by an experiment using a quartz cell of 20 cm path length containing deaerated 10^{-4} mol dm^{-3} aqueous KI. After the flash, a transient absorption was seen at around 400 nm. This was attributed to the ion I_2^-, formed as follows:

$$\left.\begin{array}{rcl} I^- & \xrightarrow{h\nu} & I^{\cdot} + e_{aq}^- \\ I^{\cdot} + I^- & \rightarrow & I_2^- \end{array}\right\} \qquad (11.26)$$

At 400 nm, this absorption was observed to decay as follows:

$t/\mu s$	50	100	150	200	250	300	350	400
A	0.277	0.189	0.156	0.122	0.102	0.087	0.083	0.065

The I_2^- ion is expected to be removed by the process,

$$I_2^- + I_2^- \rightarrow I_3^- + I^- \qquad (11.27)$$

which would imply second order kinetics. The rate equation is for a reaction involving identical species, as in Section 4.1.1, leading to the relation:

$$\frac{1}{a-x} - \frac{1}{a} = 2k_2 t \qquad (11.28)$$

Since the concentration of I_2^-, $(a - x)$, is related to the absorbance at time t, A_t, by the equation,

$$A_t = \varepsilon(a - x)l \qquad (11.29)$$

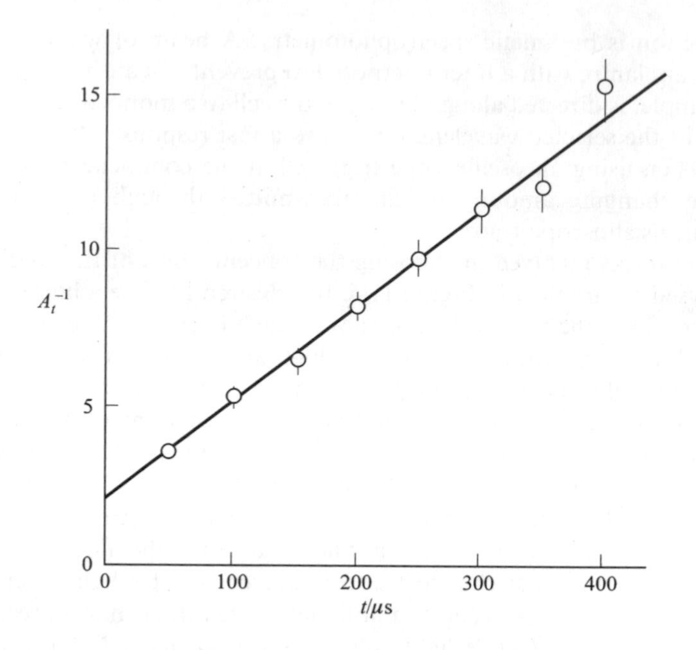

Fig. 11.7 Plot of A_t^{-1} against time for the data listed above for the decay of I_2^- ion in aqueous solution, showing the probable errors assuming that each absorbance value is accurate to ± 0.005. The optimum straight line has been calculated using equations (2.30) and (2.31).

substitution in equation (11.28) leads to the relation,

$$\frac{1}{A_t} - \frac{1}{A_o} = \frac{2k_2 t}{\varepsilon l} \tag{11.28a}$$

To test for second order kinetics, A_t^{-1} is plotted against t, as shown in Figure 11.7. The error in the reciprocal of A_t will obviously increase with t, so the best straight line is drawn using equations (2.30) and (2.31) and the slope is evaluated as $3.07 \times 10^4\ s^{-1}$.

The evaluation of a rate constant from a second order decay, unlike that from a first order decay, requires the value of the absorption coefficient. The ε value of the I_2^- ion has been measured as $1.4 \times 10^4\ dm^3\ mol^{-1}\ cm^{-1}$, so we have:

$$\begin{aligned} 2k_2 &= 3.07 \times 10^4 \times 20 \times 1.4 \times 10^4 \\ &= 8.6 \times 10^9\ dm^3\ mol^{-1}\ s^{-1} \end{aligned} \tag{11.30}$$

In more recent years, several evolutionary changes have been made to the technique of flash photolysis. The major innovation has been that, as the photolysing light, a flash lasting several tens of microseconds has been replaced by a laser pulse whose duration lies on the nanosecond or even the picosecond scale. In addition to this vastly improved time resolution, the laser pulse differs in that it is monochromatic; the wavelength used needs to be suitable for the excitation of the compound to be studied. Also, the laser light is a parallel coherent beam, so it is allowed to enter the cell at close to normal incidence. The analysing beam is then either perpendicular to this, or along an almost parallel axis, as sketched in Figure 11.8. In neither case is there a strictly uniform concentration of the transient intermediate in the

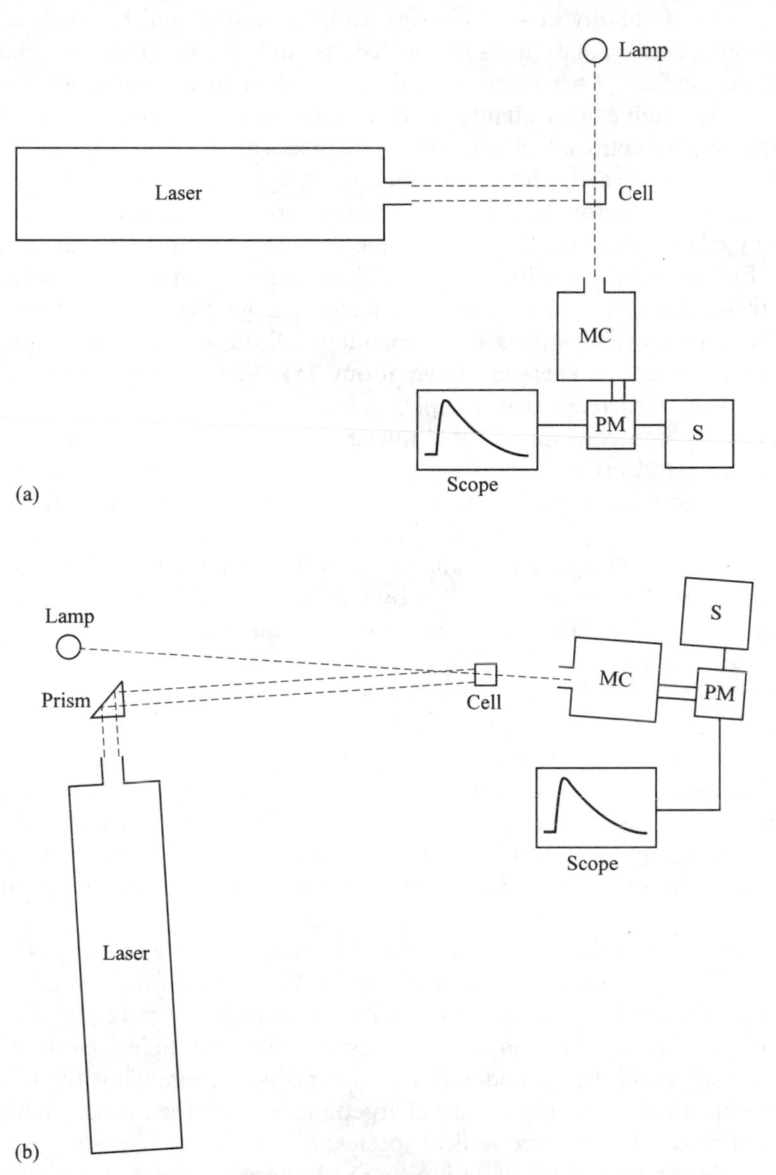

Fig. 11.8 Sketches of arrangements used in laser flash photolysis systems: (a) analysing beam perpendicular to the laser pulse; (b) analysing beam almost collinear with the laser pulse.

volume being monitored, but there are means of minimising the attendant errors.

11.4 Primary effects of ionising radiations

The rational sub-division of ionising radiations distinguishes high energy electromagnetic radiations, such as X-rays and γ-rays, from fast charged particles, such as β^--particles, α-particles, accelerated electrons, protons and deuterons. High energy quanta deposit their energy by two main processes. In the photo-electric effect, a quantum displaces an electron from its orbital, with the excess energy imparting kinetic energy to the displaced electron. The cross-section for the occurrence of this process increases the lower the energy of the photon and the higher the atomic number of the target material. In Compton scattering, the consequences are a displaced and energetic electron, along with a quantum of lesser energy than the original. The probability of this process is independent of the atomic number of the atoms involved. At energies above about 25 keV, Compton scattering is a more likely occurrence than the photoelectric effect.

Thus the consequence of these interactions of high energy quanta is to produce fast electrons, behaving similarly to β^--particles or fast electrons from an accelerator. So for all types of radiation, the loss of energy to the irradiated substance occurs ultimately by the interaction of a fast charged particle with the target atoms and molecules. At moderate or low energies, the major process of energy loss is by Coulombic interaction with the electrons of the target material, leading to electronic excitation or, more probably, to ionisation.

It was shown by Bethe that the linear rate of loss of kinetic energy of the fast charged particle is approximately proportional to the square of its charge and inversely proportional to the square of its velocity. Also, it is proportional to the electron density of the target material. This rate of loss, widely known as the *linear energy transfer* (LET) of the ionising radiation, increases along the track of any such particle so that, per unit length, there will be many more ionisations in the later than in the earlier stages of the track.

Where an electron has been displaced from a molecule by the passage of the fast charged particle, there will now be Coulombic forces of attraction between the electron and the cation, which may perhaps lead to their recombination. The resulting molecule is usually formed in an electronically excited state and it may undergo homolytic dissociation. Thus the immediate results of the passage of a fast charged particle are the formation of ions, excited molecules and free radical species.

To understand the effects of ionising radiations on gases, it is relevant to study the reactions of the ions formed from these molecules by electron impact. Some ion–molecule reactions are considered in Sections 12.4 and 12.5, principally from kinetic and mechanistic standpoints.

The major objective for the remainder of this chapter is to illustrate some ways in which radiation chemical techniques can be used to study the kinetics of chemical reactions. Many pertinent examples of the principles involved are available from radiation chemical studies of aqueous systems. Since biological systems are invariably water-based, this choice may be claimed to be apposite as well as convenient.

11.5 Radiation chemistry of aqueous solutions

A clear distinction may be made between photochemistry in solution and radiation chemistry in solution. With the former, one is almost invariably using light to which the solvent is transparent, absorbed only by the solute. With ionising radiation, it is impossible to have absorption solely by a solute. In a chemical system of two components, the energy distribution between components will be on the basis of their relative electron densities. Thus, when a dilute solution in any liquid is γ-irradiated, the energy will be almost exclusively absorbed by the liquid. This is known as the phenomenon of indirect action, and it explains why studies of the effect of ionising radiations on dilute aqueous solutions have been used to investigate the radiation chemistry of water.

Early studies in acidified solutions led to the postulate known as the *Working Hypothesis,* which is that the eventual chemical effects of irradiation may be comprehended on the assumption that within *ca* 10^{-8} s of the passage of the fast charged particle, H atoms, OH radicals and molecules of H_2 and of H_2O_2 have been formed, in yields which are fixed for any single type of radiation. That is,

$$H_2O \rightsquigarrow G_H \cdot H^{\cdot} + G_{OH} \cdot \dot{O}H + G_{H_2} \cdot H_2 + G_{H_2O_2} \cdot H_2O_2 \qquad (11.31)$$

where G_X denotes the yield of radicals/molecules of X per 100 eV energy input. The symbol "$\rightsquigarrow$" is used to indicate a process brought about by ionising radiation.

The operation of this hypothesis may be demonstrated in regard to the irradiation of dilute solutions of ferrous sulphate in dilute sulphuric acid. In a deaerated solution, the following reactions occur:

$$\dot{O}H + Fe^{2+} \rightarrow OH^- + Fe^{3+} \qquad (11.32)$$

$$H_2O + H^{\cdot} + Fe^{2+} \rightarrow H_2 + Fe^{3+} + OH^- \qquad (11.33)$$

$$H_2O_2 + Fe^{2+} \rightarrow OH^- + \dot{O}H + Fe^{3+} \qquad (11.34)$$

The yields of two products, namely of Fe^{3+} and of H_2, are readily measured in this system. For γ-radiolysis, $G(Fe^{3+}) = 8.2$ and $G(H_2) = 4.05$. From the Working Hypothesis and reactions (11.32)–(11.34), we would deduce as follows:

$$G(Fe^{3+})_{deaer} = G_{OH} + G_H + 2G_{H_2O_2} \qquad (11.35)$$

$$G(H_2)_{deaer} = G_{H_2} + G_H \tag{11.36}$$

If the solution is not deaerated, only reaction (11.33) of the above mechanism is affected. The H atom now reacts with O_2 and this leads to the oxidation of three molecules of Fe^{2+} in reactions (11.38), (11.34) and (11.32):

$$H^{\cdot} + O_2 \rightarrow H\dot{O}_2 \tag{11.37}$$

$$H\dot{O}_2 + Fe^{2+} \rightarrow Fe^{3+} + HO_2^- (\xrightarrow{H^+} H_2O_2) \tag{11.38}$$

Consequently, in aerated solution, the yield of Fe^{3+}, which may be measured as $G(Fe^{3+}) = 15.5$, is given by:

$$G(Fe^{3+})_{aer} = G_{OH} + 3G_H + 2G_{H_2O_2} \tag{11.39}$$

Subtracting equation (11.35) from this leads to the result that $G_H = 3.65$. From equation (11.36) we can then deduce that $G_{H_2} = 0.4$. From studies in other systems it may be shown that $G_{H_2O_2} = 0.8$ and $G_{OH} = 2.9$, and equation (11.35) is consistent with these values.

A solution containing 10^{-3} mol dm^{-3} ferrous sulphate in 0.4 mol dm^{-3} H_2SO_4 is the most widely used dosimeter in radiation chemistry. It is called after Fricke, one of the pioneers of the subject, and has most of the desirable attributes. It is cheap, easily obtainable, reliable and easy to use. The response is linear with dose and is easily measured by UV spectrophotometry, since ferric ion absorbs strongly at 305 and 224 nm, where ferrous ion does not. The slight variation of $G(Fe^{3+})$ with the type of radiation is in no sense a shortcoming.

In neutral solution, the radical species formed on the radiolysis of water, in addition to the $\dot{O}H$ radical, is the solvated electron. The origin of this species can be understood from the description in Section 11.4 of the displacement of electrons by the fast charged particle. An electron that has too little energy to cause further ionisation or excitation (i.e. a sub-excitation electron) will lose energy chiefly by interacting with molecular dipoles, before being slowed to thermal energy. In a polar substance such as water there is then a very high probability that it will become trapped and solvated and so will not be drawn back to the cationic residue of its parent molecule.

The concept of the solvated electron was first invoked by Kraus in 1907 to explain the very high electrical conductivity and low density of the blue solutions of alkali metals in liquid ammonia:

$$Na \rightleftharpoons Na^+ + e^- \tag{11.40}$$

However, whereas the ammoniated electron is stable with respect to reaction with solvent, the hydrated electron undergoes a bimolecular reaction to yield H_2:

$$e_{aq}^- + e_{aq}^- + 2H_2O \rightarrow H_2 + 2OH^- \tag{11.41}$$

Thus it is an unstable species, stoichiometrically equivalent to the H atom, which is its conjugate acid:

$$H \rightleftharpoons H^+ + e^-_{aq} \qquad (11.42)$$

In strongly acid solutions, a thermalised electron may be expected to react so rapidly with H^+ that it would scarcely have time to become properly solvated, so the fleeting existence of the solvated electron in acid solution is ignored in equation (11.31). In neutral solution, the Working Hypothesis requires not only the presence of the solvated electron, but also slightly different yields per 100 eV of energy input:

$$H_2O \rightsquigarrow 2.85e^-_{aq} + 2.85\,\dot{O}H + 0.6\,H^{\cdot} + 0.45\,H_2 + 0.75\,H_2O_2 \qquad (11.43)$$

Since the γ-radiolysis of water readily generates reactive free radicals in good yield, this technique lends itself to the study of competition kinetics. In neutral solution containing N_2O and nitrite ion, the hydroxyl radical and the H atom act as oxidant and reductant of nitrite ion:

$$OH + NO_2^- \rightarrow OH^- + NO_2 \qquad (11.44)$$

$$H + NO_2^- \rightarrow H^+ + NO_2^{2-} \qquad (11.45)$$

The solvated electron is the subject of competition between the two solutes:

$$\left.\begin{array}{lcl} e_{aq}^- + NO_2^- & \xrightarrow{1} & NO_2^{2-} \\ e_{aq}^- + N_2O & \xrightarrow{2} & N_2 + O^- (\xrightarrow{H^+} \dot{O}H) \end{array}\right\} \qquad (11.46)$$

Of the products formed in steps (11.44) to (11.46), N_2 is stable and is readily estimated. The others react as follows:

$$\left.\begin{array}{lcl} NO_2^{2-} + NO_2 & \rightarrow & 2NO_2^- \\ NO_2 + H_2O & = & \tfrac{1}{2}NO_2^- + \tfrac{1}{2}NO_3^- + H^+ \end{array}\right\} \qquad (11.47)$$

Since N_2 is formed from the fraction of solvated electrons reacting by step 2 rather than by step 1, we have:

$$G(N_2) = G_{e^-}\frac{k_2[N_2O]}{k_1[NO_2^-] + k_2[N_2O]} \qquad (11.48)$$

Inverting this equation, we obtain:

$$\frac{1}{G(N_2)} = \frac{1}{G_{e^-}}\left\{1 + \frac{k_1[NO_2^-]}{k_2[N_2O]}\right\} \qquad (11.49)$$

The relevant graph is of $\{G(N_2)\}^{-1}$ against $[NO_2^-]/[N_2O]$, and is closely analogous to the plot of data from the photochemical system shown in Figure 11.4. The ratio of slope to intercept yields the rate constant ratio k_1/k_2.

Another possible application of the radiolysis of water is to initiate a free radical chain reaction. As compared to thermal or photochemical initiation, the use of radiolysis has the advantage that no initiator substance need be added to the system. Moreover, a uniform rate of initiation may be achieved throughout the solution provided a penetrating type of radiation is used.

11.6 Pulse radiolysis

To a considerable extent the major aim in the early work on pulse radiolysis, the radiation chemical counterpart of flash photolysis, was to detect and thus verify the presence of transient species. For example, in 1962, Hart and Boag showed that when neutral water was pulsed with a beam of 1.8 MeV electrons, there resulted a transient absorption in the long wavelength end of the visible range. This was attributed to the solvated electron, a species which had at that time been mooted but whose existence had not been clearly established.

Pulse radiolysis has since become a most important technique for the study of the rates of many reactions. This is particularly true for reactions of the hydrated electron and of the hydroxyl radical, which have been discussed in the previous section as reactive intermediates formed in the radiolysis of aqueous solution. However, the technique may also be used to study reactions of other species which may be produced by secondary reactions of transient species formed as a consequence of the pulse of radiation.

The technique of pulse radiolysis closely resembles that of laser flash photolysis. A pulse of accelerated electrons (of around 2 or 3 MeV) is produced from an accelerator, such as a Van de Graaff or a linear accelerator. In earlier decades, these pulses might have been of microsecond duration, but more recently pulses of a few nanoseconds have been widely available. This pulse enters, at normal incidence, a piece of quartz tubing of rectangular cross-section, which serves as the reaction cell. This cell is usually part of a flow system, so that the solution may be changed without disturbing the geometry of the set-up.

Apart from some special work for which it is appropriate to monitor the reaction using conductivity, the technique almost universally employed is kinetic spectrophotometry. A major difference from flash photolysis is that whereas scattered light from a flash tube is innocuous provided the eyes are protected, it is necessary to have a substantial barrier between the pulse radiolysis cell and the operator. For various reasons, the light beam, transmitted through the cell, has to be directed, via a series of mirrors, outside the irradiation room before it enters the monochromator and then the photomultiplier. For work on short time scales, it is appropriate temporarily to boost the xenon arc lamp current to increase the light emission of the lamp by a factor of *ca* 25, as a means of improving the signal-to-noise ratio.

Initially, the major kinetic use of pulse radiolysis was the study of the rates of reaction of selected solutes with the hydrated electron. Since this species absorbs strongly in the visible and near IR, with ε_{max} at 720 nm equal to $1.85 \times 10^4\,dm^3\,mol^{-1}\,cm^{-1}$, the experiments can readily be performed under pseudo-first order conditions, with the concentration of the hydrated electron at the end of the pulse around $10^{-5}\,\mathrm{mol\,dm^{-3}}$ and much less than that of the solute, S. From the decay of the absorbance due to the electron, each experiment yields a pseudo-first order rate constant, k'. The plot of k' against the concentration of S will have a small intercept, arising from the slow decay of the electron concentration in the absence of S. This plot should be linear and its slope yields the second order rate constant for the reaction of e_{aq}^- with S.

The ease with which the traces of photomultiplier current against time can be recorded by computer and treated using an interactive program to yield k' means that such calculations are both speedy and precise. More rate constants for reactions of the hydrated electron have been measured, mostly by this technique, than of any other chemical species. In a number of instances, these rate constants are found to lie right at the upper limit for diffusion-controlled reactions, that is, in excess of $10^{10}\,\mathrm{dm^3\,mol^{-1}\,s^{-1}}$. For such fast reactions, accurate measurement is achievable using short (e.g. 25 ns) pulses of fast electrons, with concentrations of the solute in the region of 10^{-4} to $10^{-3}\,\mathrm{mol\,dm^{-3}}$, giving a half-life under or around 1 μs.

The hydrated electron may be regarded as a simple reducing species. It reacts, for example, with aqueous benzene by attachment to the π-electron system, with a rate constant of $1.0 \times 10^7\,\mathrm{dm^3\,mol^{-1}\,s^{-1}}$. The effect on the rate constant of introducing a substituent to the benzene ring is shown in Figure 11.9. While the rate rises with increase in the Hammett substituent constant, σ, apparently with a reaction constant $\rho = +4.8$, the scatter in the points must raise doubts as to whether the only relevant parameter is the electron-withdrawing or -donating character of the substituent. The halogen atoms all have comparable values of σ, but the hydrated electron reacts more quickly with iodobenzene than with fluorobenzene by a factor in excess of 10^2.

Other studies have been directed towards measuring the rate constant for the reaction

$$e_{aq}^- \quad \rightarrow \quad H^\cdot + OH^- \tag{11.50}$$

for which the half-life, in carefully purified water, made slightly alkaline by the addition of NaOH, was found to be 660 μs. Baxendale realised that this parameter, in conjunction with the rate constant for the conversion of the hydrogen atom to the hydrated electron at pH values in excess of 10, leads to reliable thermodynamic data for the hydrated electron, whose reactivity precludes normal methods of study. Thus the standard Gibbs energy of hydration of the electron is known to be $-157\,\mathrm{kJ\,mol^{-1}}$, whereas that of the large anion, I^-, is $-238\,\mathrm{kJ\,mol^{-1}}$. This would suggest that the negative charge of the hydrated electron is not as localised as some models of this species might suggest.

The other main transient species generated in the radiolysis of water, the hydroxyl radical, absorbs only very weakly and in the lower UV region where many other substances absorb. Consequently, it is not feasible to

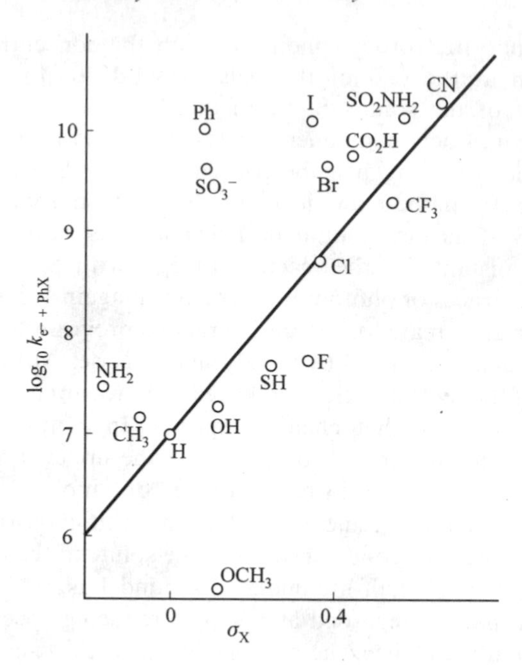

Fig. 11.9 Plot of $\log_{10}k$ for the reaction of the hydrated electron with PhX in aqueous solution, against the Hammett substituent constant, σ_X. The rate constants are taken from the compilation in Buxton, G. V., C. L. Greenstock, W. P. Helman and A. B. Ross, *J. Phys. Chem. Reference Data*, 1988, **17,** 513–886, and the σ values from McDaniel, D. H. and H. C. Brown, *J. Org. Chem.*, 1958, **23**, 420–7. The more positive of the σ values for the *meta-* and *para-* positions has been taken as the indicator of the effect on the reactivity of benzene produced by a single substituent. The line gives the Hammett reaction constant $\rho = +4.8$.

monitor the reactions of the $\dot{O}H$ radical in the manner described above for the hydrated electron. However, sometimes the reaction can be followed by observing the build-up of a reaction product.

The $\dot{O}H$ radical oxidises 3-ferrocenylpropanoate anion to the corresponding ferricenium species:

$$\begin{array}{c} OH + C_5H_4(CH_2)_2CO_2^- \\ | \\ Fe \\ | \\ C_5H_5 \end{array} \longrightarrow \begin{array}{c} OH^- + C_5H_4(CH_2)_2CO_2^- \\ | \\ Fe^+ \\ | \\ C_5H_5 \end{array} \qquad (11.51)$$

It is convenient to study this reaction in a solution saturated with N_2O, so that, by the occurrence of the second part of reaction (11.46), the hydrated electron is quantitatively converted into the hydroxyl radical. The build-up of the ferricenium zwitterion may be followed using its absorption peak at 625 nm. Since this reaction also is studied under pseudo-first conditions, each experiment yields a pseudo-first order rate constant, k', and the second order rate constant for reaction (11.51) is obtained from the slope of the plot of k' against the concentration of the ferrocenyl anion.

Strictly speaking, the previous statement is not entirely true. There is evidence that, to a minor extent, the OH radical reacts with this anion by H atom abstraction. In the light of the discussion in Chapter 3 on parallel reactions and equation (3.32), the second order rate constant obtained as described in the previous paragraph is the aggregate rate constant for all the reactions of $\dot{O}H$ with ferrocenylpropanoate, and this was reported as $1.1 \times 10^{10}\,dm^3\,mol^{-1}\,s^{-1}$.

Where the OH radical reacts with a solute S_1, to yield a product, X, which does not absorb within a convenient wavelength range, it may be possible to measure the rate constant for this reaction by pulse radiolysis using a competition method. To achieve this, a reference solute S_2 is added, which on reaction with OH produces a strongly absorbing product, Y:

$$\left.\begin{array}{lcl} OH + S_1 & \xrightarrow{1} & X \\ OH + S_2 & \xrightarrow{2} & Y \end{array}\right\} \qquad (11.52)$$

The fraction of the OH radicals which reacts with S_2 is given by $k_2[S_2]/\{k_1[S_1] + k_2[S_2]\}$, which should also represent the ratio of the absorbance increase, A, achieved with both S_1 and S_2 present to the value, A_o, with only the reference solute present. Thus we have:

$$\begin{aligned} \frac{A_o}{A} &= \frac{k_1[S_1] + k_2[S_2]}{k_2[S_2]} \\ &= 1 + \frac{k_1[S_1]}{k_2[S_2]} \end{aligned} \qquad (11.53)$$

This means a plot of A_o/A against $[S_1]/[S_2]$ should yield a linear plot with an intercept of unity and a slope of k_1/k_2. The rate constant k_2 may be accessible by direct determination, thus enabling the evaluation of k_1.

Pulse radiolysis may also be used to measure the rate constants for reactions of secondary species. For example, in the presence of thiocyanate ion we may have the reactions,

$$\left.\begin{array}{lcl} OH + SCN^- & \rightarrow & OH^- + SCN^{\cdot} \\ SCN^{\cdot} + SCN^- & \rightarrow & (SCN)_2^- \end{array}\right\} \qquad (11.54)$$

where the ion $(SCN)_2^-$ absorbs strongly at 480 nm. If another solute, S, is present in much lower concentration than thiocyanate, it will not materially interfere with reaction (11.54), but the rate of the reaction of $(SCN)_2^-$ with it may be deduced by monitoring at 480 nm. The decay of $(SCN)_2^-$ ion will be pseudo-first order, and the rate constant for the reaction with **S** will be found from the slope of the plot of this rate constant against the concentration of **S**.

Suggested reading

Gilbert, A. and J. E. Baggott, 1991. *Essentials of Molecular Photochemistry*. Oxford: Blackwell.

Simons, J. P., 1971. *Photochemistry and Spectroscopy*. London: Wiley-Interscience.

Arnold, D. R., N. C. Baird, J. R. Bolton, J. C. D. Brand, P. M. W. Jacobs, P. de Mayo and W. R. Ware, 1974. *Photochemistry: An Introduction*. New York: Academic.

Wayne, R. P., 1988. *Principles and Applications of Photochemistry*. London: Oxford University Press.

Rehm, D. and A. Weller, 1970. Kinetics of fluorescence quenching by electron and H atom transfer. *Israel J. Chem.*, **8,** 259–71.

Logan, S. R., 1990. Spatial inhomogeneity effects in photochemical kinetics. *J. Chem. Educ.*, **67**, 872–5.

Porter, G. and M. A. West, 1974. Flash photolysis. In *Investigations of Rates and Mechanisms of Reactions*, 3rd edition, Part II (ed. Hammes, G. G.) pp. 367–462. New York: Wiley-Interscience.

Spinks, J. W. T. and R. J. Woods, 1990. *An Introduction to Radiation Chemistry*, 3rd edition. New York: Wiley.

Dorfman, L. M., 1974. Pulse radiolysis. In *Investigations of Rates and Mechanisms of Reactions,* 3rd edition, Part II (ed. Hammes, G. G.), pp. 463–519. New York: Wiley-Interscience.

Bensasson, R. V., E. J. Land and T. G. Truscott, 1983. *Flash Photolysis and Pulse Radiolysis. Contributions to the Chemistry of Biology and Medicine*. Oxford: Pergamon.

Baxendale, J. H. and M. A. J. Rodgers, 1978. Contributions of pulse radiolysis to chemistry. *Chemical Society Reviews*, **7**, 235–63.

Problems

11.1 In a solvent of viscosity $\eta = 3.26 \times 10^{-4}\,\mathrm{kg\,m^{-1}\,s^{-1}}$, the fluorescence from an aromatic hydrocarbon was found to be quenched by added solute Q, at 293 K, as follows:

$[Q]/10^{-3}\,\mathrm{mol\,dm^{-3}}$	0	0.5	1.0	1.5	2.0
I_f/I_f^o	1.00	0.67	0.49	0.40	0.33

On the assumption that the quenching process is diffusion controlled, determine the mean lifetime of the excited singlet state which emits the fluorescence.

11.2 Under UV light, hydroxide ion reacts with *m*-nitroanisole (**I**) to give *m*-nitrophenol (**II**):

$$\mathbf{I}\ (m\text{-}NO_2C_6H_4OMe) + OH^- \xrightarrow{h\nu} \mathbf{II}\ (m\text{-}NO_2C_6H_4OH) + MeO^-$$

The addition of a triplet quencher, Q, causes the quantum yield ϕ of (**II**) to be decreased, as follows:

$[Q]/10^{-3}\,mol\,dm^{-3}$	0	1.7	3.3	5.0	6.9	9.8
ϕ/mol Einstein^{-1}	0.37	0.32	0.28	0.25	0.22	0.18

Show that these data, obtained with OH^- concentration kept constant, are consistent with the UV excitation of (**I**) to an excited singlet state, leading, by intersystem crossing, to a triplet state with four possible fates:

(i) radiationless decay to the ground state,
(ii) attack by OH^- leading to the product, **II**,
(iii) attack by OH^- leading to the ground state,
(iv) reaction with the triplet quencher Q leading to the ground state.

(Data from den Heiger, J., T. Spee, G. P. de Gunst and J. Cornelisse, *Tet. Lett.*, 1973, **15**, 1261–4.)

11.3 A deaerated solution with the following concentrations, $[I_3^-] = 0.00197\,mol\,dm^{-3}$, $[I^-] = 0.0524\,mol\,dm^{-3}$ and $[N_2O] = 6.00 \times 10^{-3}\,mol\,dm^{-3}$, was irradiated for 30 minutes by a low pressure mercury lamp, and the yield of N_2 found to be 2.45 μmoles. Actinometry established the flux of 254 nm light as 2.55×10^{-8} Einstein s^{-1}. Given that, at 254 nm, $\varepsilon_{I^-} = 220\,dm^3\,mol^{-1}\,cm^{-1}$ and $\varepsilon_{I_3^-} = 3.3 \times 10^3\,dm^3\,mol^{-1}\,cm^{-1}$, the absorption of light by I_3^- is described by the sequence:

$$I_3^- \xrightarrow{h\nu} I_2^- + I$$
$$I + I^- \rightarrow I_2^-$$
$$I_2^- + I_2^- \rightarrow I_3^- + I^-$$

whereas for I^- we have:

$$I^- \xrightarrow{h\nu} I + e^- \qquad (\phi = 0.235)$$

evaluate the relative rate constant, $k(e^- + I_3^-)/k(e^- + N_2O)$.

11.4 A fluorimeter cell, 1 cm square, contains a solution of I_2 in *n*-hexane at 293 K. A laser pulse of 530 nm light is incident upon one face of the cell, while analysing light from a flash lamp passes across the cell, close to this cell face, and to a monochromator set at 300 nm, to a photomultiplier unit and the signal goes to an oscilloscope. The sweep was 200 ns per division and in transferring the photo of the trace to graph paper, one division became 2.34 cm.

The laser pulse causes a transient decrease in transmittance at 300 nm, attributable to a contact charge-transfer interaction between the I atom and the solvent. Starting at an arbitrary position just after the end of the laser pulse, the trace on the graph paper gives the following values for the relative intensity of the transmitted light:

Horizonal position/cm	0	1	2	3	4	5	6	7	8	9
I/I_o	0.544	0.669	0.750	0.801	0.831	0.856	0.869	0.882	0.894	0.903

Using these data, make such deductions as are possible regarding the kinetics of the reaction, $I + I \rightarrow I_2$, in *n*-hexane, whose viscosity $\eta = 3.26 \times 10^{-4}\ \mathrm{kg\,m^{-1}\,s^{-1}}$.

11.5 The following data were obtained for the γ-radiolysis of solutions containing 0.050 mol dm^{-3} NO_2^- and varying concentrations of N_2O.

$[N_2O]/\mathrm{mol\,dm^{-3}}$	0.0254	0.0225	0.0164	0.0125	0.0105
$G(N_2)$	1.59	1.47	1.18	0.99	0.86

Assuming mechanism (11.46), evaluate $k(e^- + NO_2^-)/k(e^- + N_2O)$.

11.6 Pulse radiolysis studies were carried out on solutions of $Fc(CH_2)_2CO_2^-$ that had been partially oxidised to the dipolar ion, $Fc^+(CH_2)_2CO_2^-$, where $Fc \equiv$ ferrocenyl, $C_5H_5.Fe.C_5H_4^{\cdot}$. The transient at 775 nm due to the hydrated electron was found to decay by first order kinetics. Just after pulsing, the absorbance of the solution was measured at 625 nm in a 1 cm cell.

A at 625 nm	0.010	0.095	0.184	0.247
$k'/10^7\ \mathrm{s^{-1}}$	0.11	0.94	1.82	2.37

From these data, evaluate the rate constant for the reaction of the hydrated electron with the ferricenium dipolar ion, taking its molar decadic absorption coefficient, ε, at 625 nm as 332 dm^3 mol^{-1} cm^{-1}.

11.7 A solution of 0.4 mol dm^{-3} sulphuric acid contains amounts of Fe^{2+} ion and of O_2 such that some H atoms generated by γ-radiolysis react by step (11.33) and some by step (11.37). Derive an equation for $G(Fe^{3+})$ in this regime and show how experimental data on the yield of Fe^{3+} might be used to evaluate the relative rate constant for these two reactions.

12 Reaction dynamics

Beams of atoms and molecules, effusing from an orifice in an oven into an evacuated chamber, were used to verify the Maxwell–Boltzmann law of the distribution of molecular velocities. They also enabled Stern and Gerlach to carry out the experiment with an inhomogeneous magnetic field which demonstrated the existence of quantised electron spin.

In the 1950s, there came the application of crossed molecular beams of two different chemical species to study their reaction in a more detailed way. This approach attempted to study chemical reactions under conditions where various parameters of the reacting species were rigidly controlled, so that, for example, the angle and the velocities of approach could be varied independently. The range of reactions which may be scrutinised in this way is now, due to various technical advances, much less restricted, but adequate sensitivity is not always easily achieved. Successful endeavours in this field have led to the emergence of the topic of chemical reaction dynamics.

During the last 40 years there has also been much interest in the reactions between ions and molecules, some of which exhibit very high rate constants. The detailed mechanisms can be probed in a range of ways, using equipment ranging from a slightly modified mass spectrometer to a highly sophisticated system with crossed ion and molecular beams.

12.1 Introduction

In regard to the progress of chemical reactions, the term "dynamics" was formerly used as a synonym for "kinetics". Thus van't Hoff's treatise was entitled, "Etudes de dynamique chimique". While the use of these terms in physical mechanics might suggest that kinetics is a sub-set of dynamics, current usage in regard to chemical reactions is the reverse of this: the term chemical dynamics is now used only in a restricted sense.

In Chapter 3, a broad sub-division of chemical reactions was made into those that are elementary processes, as distinct from those that take place in a series of discrete steps. The second half of this book has explored the various ways in which elementary steps may be combined in achieving the progress of a chemical reaction, so it is appropriate to conclude with a review of more detailed studies of elementary bimolecular processes.

A truly elementary reaction, "involving precisely those species appearing on the left-hand side of the stoichiometric equation", is, arguably, possible only in the gas phase, with no other molecules, whether of solvent or otherwise, capable of causing interference. So far we have considered such reactions only in a macroscopic manner, looking at the progress of a reaction involving some significant amount of substance and characterising the rapidity in terms of the appropriate rate constant. The complement to this approach is to examine, as far as it is possible, the interactions of individual pairs of molecules. More detailed questions may then be contemplated, relating both to the collisions which result in chemical reaction and to those which do not. For reactions which are considerably exothermal, one aspect is the manner in which this excess energy is partitioned between the internal modes of the products and their translational energy. The significance of chemical reaction dynamics was recognised by the award of the 1986 Nobel Prize in Chemistry to Herschbach, Lee and J. C. Polanyi.

12.2 Studies using crossed molecular beams

The technically difficult feat of observing chemical reactions in crossed molecular beams was first achieved in 1954. At a pressure of under 10^{-4} torr, the mean free path of a molecule is much greater than the dimensions of a medium sized vessel, so that a beam of molecules issuing from a slit, in the manner of the Stern–Gerlach experiment, will make scarcely any collisions with other molecules. When such a molecular beam is directed across the path of another beam, only an extremely small minority of the molecules of either will engage in collision. Particularly sensitive techniques are needed to detect the molecules of any ensuing product and to evaluate the angular distribution of the products. To facilitate this aspect, a much lower pressure is necessary than that dictated by mean free path considerations. In some systems, a value as low as 10^{-14} torr has been achieved.

Among the reactions most investigated by crossed molecular beams have been processes such as:

$$M + X_2 \quad \rightarrow \quad MX + X^{\cdot} \tag{12.1}$$

$$M + HX \quad \rightarrow \quad MX + H^{\cdot} \tag{12.2}$$

$$M + RX \quad \rightarrow \quad MX + R^{\cdot} \tag{12.3}$$

where M represents an atom of an alkali metal, X a halogen and R is usually an alkyl group. In part this is because the species M and MX can be quite readily detected using surface ionisation techniques with an efficiency of unity. An additional reason is that the reaction efficiencies are high.

It is desirable that each of the molecular beams should consist of clearly defined entities. Velocity selection has been used in many experiments. The beam emerging by effusion through a hole in the wall of an oven is made to pass through the holes in staggered slotted discs, rotating on a common shaft. However, this cuts down the beam intensity to a far greater extent than would seem to be implied by the width of the selected range of velocities, in that not all molecules within this velocity range can be transmitted. Rotational state selection is also feasible, but not the selection of the vibrational state, which is usually a more significant parameter.

If a sticky collision should occur between two species from the respective molecular beams, the resulting complex would travel in a direction and at a velocity uniquely determined by the masses and the velocities of the two colliding species. In effect, the centre of mass of these two species moves with a constant velocity before their collision and the complex would continue to move along this line. This provides the framework relevant for interpreting the angular distribution of the products, should direct reaction occur. A product may be preferentially scattered towards the direction of motion of either reactant. If towards the direction of the metal atom, this is described as forward scattering, whereas to the other side of the centre of mass it is described as backward scattering, as is illustrated in Figure 12.1.

One of the reactions that has been studied using crossed molecular beams is the process:

$$K + Br_2 \rightarrow KBr + Br \tag{12.4}$$

Much earlier work by M. Polanyi using flames suggested that this type of reaction had an unusually large cross-section. Experimentally, it was found that this exceeded 100 Å^2 and that the product molecule HBr was predominantly forward scattered.

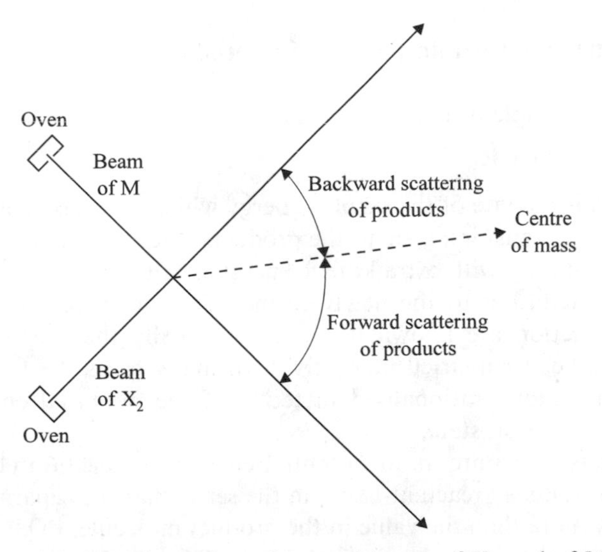

Fig. 12.1 Illustration of crossed beams of X_2 and of M, and of the designations of the preferred directions of the product, MX.

A theoretical study in 1940 led to a formulation of the mechanism involved. This is that as the K atom approaches the halogen, the low ionisation potential of the former combined with the high electron affinity of the halogen causes an electron to be transferred, yielding K^+ and Br^-Br. Thereafter, the coulombic attraction of the two ions becomes decisive and the Br^- ion is removed by K^+. This has been called the harpoon mechanism, on account of the electron transfer which, it would appear, becomes possible at a separation of nearly 6 Å. (For comparison, the sum of the Van der Waals radii of the two reactants in (12.4) is *ca* 4.5 Å.) Since the other Br atom is a virtual on-looker in the events which follow, this has also been called a spectator stripping mechanism.

By contrast, the reaction,

$$K + CH_3I \rightarrow KI + CH_3^{\cdot} \tag{12.5}$$

has a much lower cross-section of *ca* 10 $Å^2$ and it is observed that the product molecules undergo backward scattering. This is known as a recoil mechanism.

A different type of behaviour is found in reactions such as:

$$Cs + RbCl \rightarrow CsCl + Rb \tag{12.6}$$

The angular distribution of the molecular product follows neither of the previous patterns, but is located symmetrically around the centre of mass. This is consistent with the reactants forming a collision complex which survives for several rotational periods. The angular distribution of the eventual products when this complex dissociates would then be a consequence of their mutual recoil, with equal and opposite momenta.

12.3 Energy distribution within the reaction products

For a reaction of the simple displacement type,

$$P + QR \rightarrow PQ + R \tag{12.7}$$

a pertinent question is the fate of the surplus energy which is released as the system passes from the transition state to the products. Clearly, it may go to provide the product species with extra kinetic energy or alternatively it may supply vibrational excitation to the newly formed PQ molecule. Both of these patterns of reaction are known, though realistically they represent the two extremes rather than strict alternatives. In many cases the fate of the excess energy is readily rationalised in terms of the potential energy profile around the transition state.

Figure 12.2 depicts the contours of potential energy for reaction (12.7) where the transition state is "reactant-like", in the sense that the separation P–Q is substantially more than the value in the product molecule, PQ, while that of Q–R is close to that in the reactant molecule, QR. As the dotted line indicates, if the reactants approach in such an orientation and with sufficient

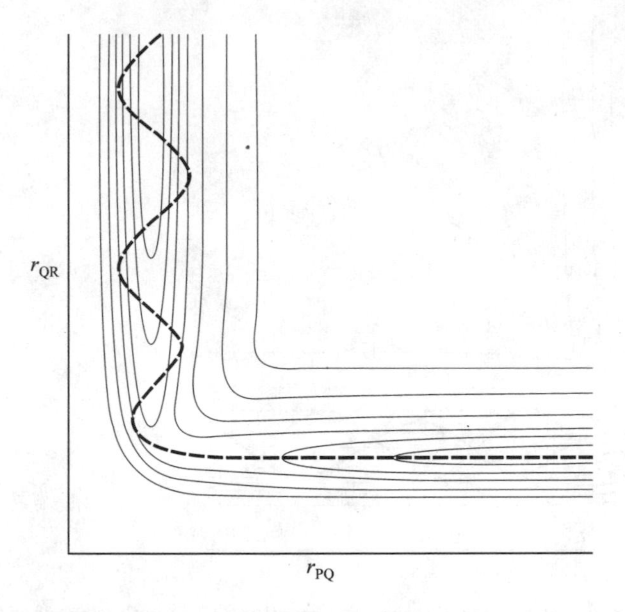

Fig. 12.2 Contours of potential energy for reaction (12.7) with a reactant-like configuration at the transition state. The dotted line indicates a probable trajectory.

relative kinetic energy that the transition state is formed, then the system may be expected to pass over the col to the products in such a way that there will be appreciable excitation in the P–Q bond.

Alternatively, the transition state may be "product-like" as illustrated in Figure 12.3. Because of the shape of the contour lines, the reactants may approach with sufficient kinetic energy but with a low probability that the transition state may be formed. However, if the reactant QR is vibrationally excited to a sufficient extent, the probability of attaining the transition state may be quite high. Furthermore, as the system passes over the col of potential energy to form the reaction products, it will do so in a manner which converts almost all the potential energy of the transition state into kinetic energy of the products, with scarcely any vibrational excitation of PQ.

A reaction which is of some interest in this context is that of the F atom with D_2:

$$F + D_2 \rightarrow DF + D \qquad (12.8)$$

This process is exoergic by $144\,kJ\,mol^{-1}$, reflecting the high reactivity of the F atom. A crossed molecular beam study of this reaction produced the results shown in Figure 12.4. This demonstrates that the DF molecular product, which undergoes backward scattering, apparently generates several cones around the centre of mass axis, reflecting different possible values of the recoil energy of DF and the D atom. The explanation is that the DF molecule is vibrationally excited to a substantial and variable extent, so that

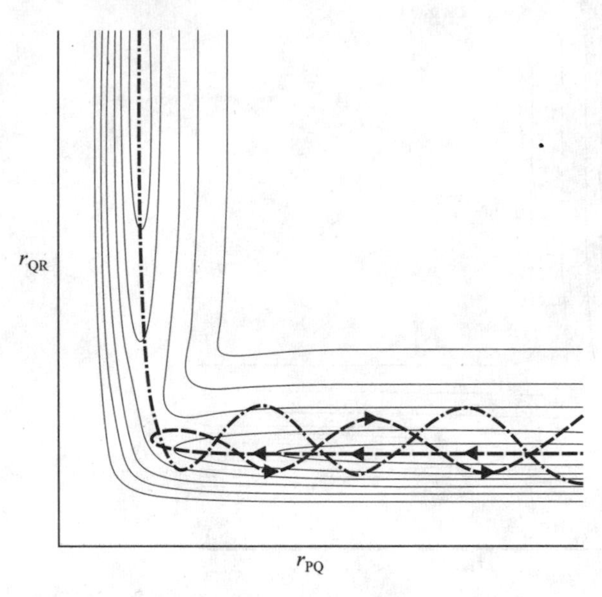

Fig. 12.3 Contours of potential energy for reaction (12.7) with a product-like configuration at the transition state. The trajectory (- - - - -) indicates the fate when QR has little vibrational excitation, whereas the other (- · · - · · -) illustrates the consequences of QR having high vibrational excitation.

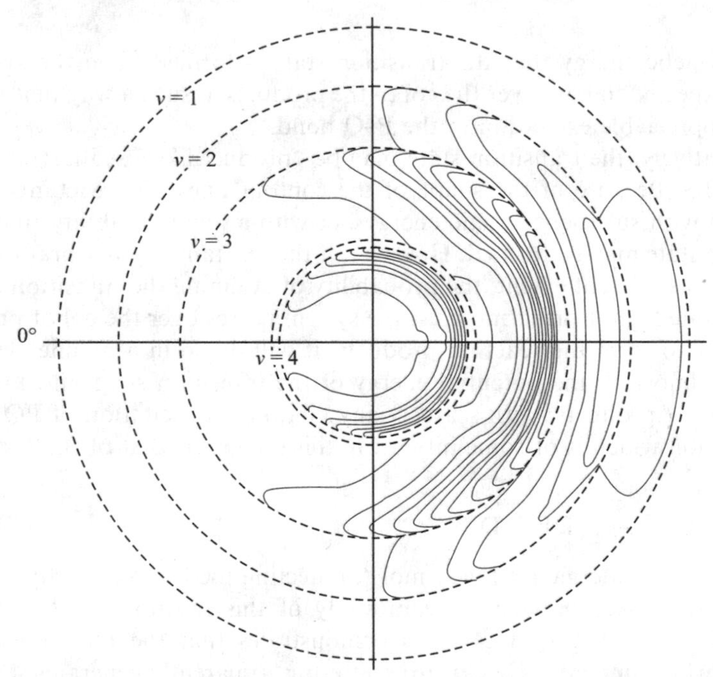

Fig. 12.4 Centre of mass contour map for the F + D_2 reaction at E_T = 7.6 kJ mol^{-1}. The dashed lines mark the maximum possible deviation of the DF molecules from the origin, for v = 1, 2, 3 and 4. (Reproduced, with permission, from Newmark, DM, AM Wodtke, GN Robinson, CC Hayden, K Shobatke, RK Sparks, TP Schafer and YT Lee, *J. Chem. Phys.*, 1985, **82**, 3067–77).

the amount of excess energy available to go into the relative kinetic energy of the products may have one of several values. Thus the concentric rings shown in Figure 12.4 correspond, commencing from the outside, to increasing values of the vibrational quantum number, v.

Ab initio calculations of the potential energy surface for the F + D_2 system show, not surprisingly, an early saddle point, before the D–D separation has increased significantly, so that on theoretical grounds this reaction is analogous to that shown in Figure 12.2.

Infra-red chemiluminescence techniques pioneered by J. C. Polanyi are capable of tackling reactions of this nature in a different way. In the Measured Relaxation (MR) method, a flow system reaction vessel has three sapphire windows to permit IR emissions to be detected. On separately adding the reactants upstream of the first viewing point, the IR emissions from the vibrationally excited product molecules may be observed at all three positions. Back-extrapolation of the respective intensities to the point of mixing then yields the relative amounts of the product molecule in the various vibrational states.

The results from these two techniques show close, though not complete, agreement. For the products of reaction (12.8), the populations of the vibrational states $v = 1, 2, 3$ and 4, as measured by crossed molecular beams, are 0.08, 0.30, 0.44 and 0.18. The corresponding fractions as obtained from chemiluminescence measurements are 0.08, 0.22, 0.44 and 0.26, which shows concurrence that the most highly populated vibrational state of the newly formed DF molecule is that with $v = 3$.

12.4 Ion–molecule reactions

Most of the early studies were carried out using a mass spectrometer, but operating it under conditions quite different from those used for analytical work. For the latter, it is desirable to have a sufficiently low pressure in the ion source that an ion is quite unlikely to collide with any molecule before it leaves the ion source. Thus the ions detected after electromagnetic separation will be precisely those generated in the original ionisation event. If this is achieved by electron impact, the abundance of each ion will depend on the electron-accelerating voltage, E, since there is an appearance potential below which it is not formed.

As the gas pressure in the ion source is increased, the stage is reached where the mean free path becomes comparable to the distance from the electron beam to the exit slit. It is then feasible that an ion may undergo a collision with a gas molecule within the ion source. Should reaction occur, it will be the product ion which is subsequently detected.

In 1940, it was found that when water vapour was admitted to the ion source of a mass spectrometer, the ions detected included one at $m/z = 19$, corresponding to the ion H_3O^+. The size of the peak was proportional to the square of the water vapour pressure and the appearance potential was

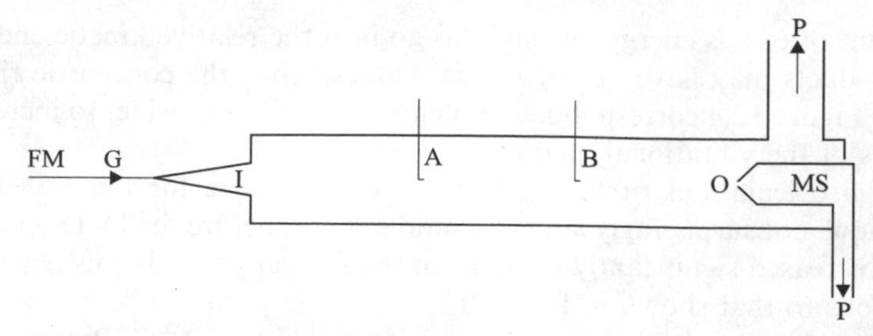

Fig. 12.5 Sketch of the drift tube of a flowing afterglow system. G, gas inlet; FM, flow meter; I, ioniser; A,B, alternative points for addition of a reactant; O, orifice to mass spectrometer (MS); P, pumps.

the same as for $m/z = 18$, for the ion H_2O^+. These facts indicated that H_3O^+ was a secondary ion, formed in the process:

$$H_2O^+ + H_2O \quad \rightarrow \quad H_3O^+ + OH \qquad (12.9)$$

Another technique used for the study of ion-molecule reactions is the drift tube, of which an outline is shown in Figure 12.5. One advantage of this method is that, during the period when reaction is occurring, the environment of the ion is more carefully controlled in regard to electric field gradients. Also, with the drift tube it is much more feasible to form ions from one molecule and then look at their reaction with a totally different species.

The rate constants of ion-molecule reactions are usually high and some are extremely high. Values have been reported in excess of $10^{-9}\,cm^3\,molec^{-1}\,s^{-1}$, reflecting the fact that for gas phase studies a convenient unit of concentration is 1 molecule per cm^3. Multiplying by $N_A/10^3$ converts the second order rate constant to the units more often used in solution and yields $6.0 \times 10^{11}\,dm^3\,mol^{-1}\,s^{-1}$, which is well in excess of the normal upper limit for a reaction between two neutral molecules in the gas phase.

The explanation for this unusually high value is that as an ion approaches an isotropic molecule it causes a dipole to be induced in the latter. The consequent attraction between ion and induced dipole (or, in the case of an anisotropic molecule, a permanent dipole) means that the effective cross-section for their collision is greater than for two uncharged species of precisely the same dimensions. An illustration of the relative motion of ion and molecule is shown in Figure 12.6.

Known ion–molecule reactions may be categorised as follows.

12.4.1 Proton transfer reactions

An example of this is the process,

$$C_2H_6^+ + D_2O \quad \rightarrow \quad HD_2O^+ + C_2H_5 \qquad (12.10)$$

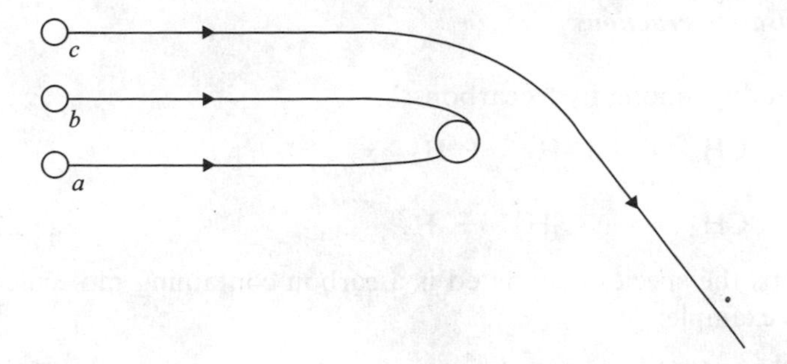

Fig. 12.6 Illustration of the interactions between an ion and a polarisable molecule. Trajectory *a* would lead to a collision without any attractive forces. Trajectory *b* leads to a collision only because of the mutual attraction. Trajectory *c* demonstrates the effects of attractive forces even where no collision occurs.

which has a rate constant of $7.8 \times 10^{11}\,dm^3\,mol^{-1}\,s^{-1}$. The highest known rate constants come from reactions in this category. For some reactions, such as,

$$CH_4^+ + CH_4 \rightarrow CH_5^+ + CH_3^{\cdot} \tag{12.11}$$

$$H_2O^+ + H_2O \rightarrow H_3O^+ + OH \tag{12.12}$$

it is not immediately clear whether the product ion is formed by the transfer of a proton or by the transfer of a hydrogen atom. However, this question can be answered using the drift tube, illustrated in Figure 12.5. By adding the normal molecule at G to produce the primary ion and the fully deuterated molecule at A to react with it, a situation is achieved where the alternative mechanisms lead to different products. Usually it is found that both processes occur but that proton transfer predominates, e.g.

$$\begin{aligned} CH_4^+ + CD_4 &\rightarrow \text{(mostly) } CH_3 + CHD_4^+ \\ &\rightarrow \text{(a little) } CH_4D^+ + CD_3^{\cdot} \end{aligned} \tag{12.13}$$

12.4.2 Neutral atom transfer

Some examples of H atom transfer are more explicit, for example:

$$Ar^+ + H_2 \rightarrow ArH^+ + H^{\cdot} \tag{12.14}$$

The transfer of larger atoms may also occur, as in the processes,

$$O^+ + N_2 \rightarrow NO^+ + N^{\cdot} \tag{12.15}$$

$$CHI^+ + CH_3I \rightarrow CHI_2^+ + CH_3^{\cdot} \tag{12.16}$$

which involve the transfer of N and of I atoms.

12.4.3 Condensation reactions

These occur readily among hydrocarbons:

$$CH_3^+ + CH_4 \rightarrow C_2H_5^+ + H_2 \quad (12.17)$$

$$C_2H_5^+ + CH_4 \rightarrow C_3H_7^+ + H_2 \quad (12.18)$$

In some systems the species eliminated is a carbon-containing molecule or radical, as for example:

$$C_2H_4^+ + C_2H_4 \rightarrow C_3H_5^+ + CH_3^{\cdot} \quad (12.19)$$

$$C_3H_6^+ + C_3H_6 \rightarrow C_4H_8^+ + C_2H_4 \quad (12.20)$$

Negative ions are not formed in abundance in hydrocarbon systems but are in evidence when more electronegative atoms are present, such as N, O, S or a halogen atom. Whereas, for the formation of the parent positive ion by electron impact, it is merely necessary for the electrons to have *sufficient* energy, in order to form a negative ion, the electron must have the *correct* energy, since the reaction,

$$M + e^- \rightarrow M^- \quad (12.21)$$

is a resonance process. Negative ions exhibit much the same range of reactions as do carbocations.

12.5 Dynamics of ion–molecule reactions

In seeking to elucidate details of the progress of a reaction, ion–molecule processes possess advantages over those involving two neutral species in that there will always be one ionic product, easily detected and identified. Also, an ionic reactant is more easily presented in a controlled fashion than is a neutral species.

One technique employed in this regard was the use of a conventional mass spectrometer, but with the potentials in the source, shown in Figure 12.7, considerably modified. With the cage at +8 V with respect to the filament, the electrons emitted by the latter traverse the cage at an energy below the ionisation potential. With the trap at *ca* +20 V, the electrons are then accelerated on leaving the cage, so that primary ions may be generated close to the trap.

These ions are then subject to the potential gradient between the trap and the cage, so that positive ions are attracted towards the latter and some fraction will enter the cage, by the same orifice, with a kinetic energy of a few eV. These primary ions, travelling horizontally across the trap, are not readily deflected towards the slit and thus do not register on the ion collection system. However, it is possible for secondary ions, produced by the reaction of these ions within the cage, to reach the slit and be detected.

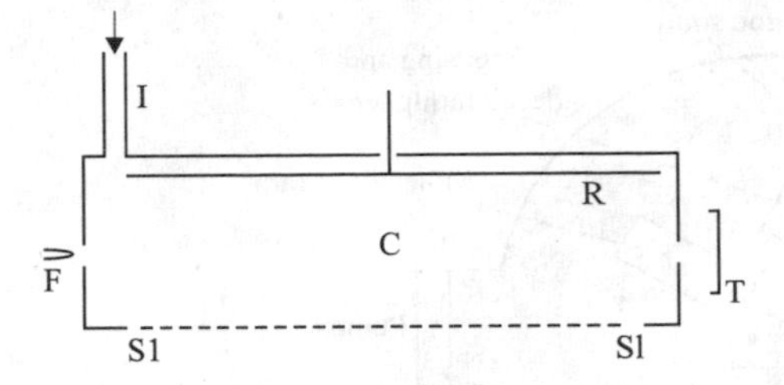

Fig. 12.7 Illustration of the ionisation chamber of a mass spectrometer, with ions generated by electron impact: F, filament; C, cage; I, gas inlet; T, trap; Sl, ion exit slit; R, ion repeller.

When methane is admitted at the inlet system of the mass spectrometer, the ion CH_5^+ is detected at a peak size proportional to the square of the methane pressure. This and the manner of the dependence of the peak on the trap voltage indicate that the secondary ion is formed by reaction (12.11), from a CH_4^+ ion produced outside the cage. If the CH_5^+ ion were formed only by H atom transfer, the product ion must inevitably retain so large a proportion of the momentum of the primary ion that it could not reach the slit. This means that a different mechanism, namely proton transfer, must occur: it does not, of course, rule out H atom transfer.

However, the significance of these experiments, which were carried out before the drift tube studies referred to in Section 12.4, is that they shed light on the nature of the proton transfer reaction. For a secondary ion to be detected in this system, it must be formed in a process in which there is only limited momentum transfer from the primary ion. This means that the reaction, between a CH_4 molecule at thermal energy and a CH_4^+ ion with at least 5 eV of kinetic energy, must occur by a proton stripping mechanism, rather than involving an intermediate complex. It may be shown that the minimum fraction of the kinetic energy of the primary ion transferred to the secondary ion is $(m/M)^2$, where M denotes the mass of the accepting molecule and m that of the entity being transferred. When CD_4 is used, the peaks due to these secondary ions are, except at very low energies, found to be appreciably smaller than for CH_4. This is entirely consistent with the proposed mechanism, since in the case of D^+ transfer to CD_4, the fraction $(m/M)^2$ is greater by a factor of 2.6 than for H^+ transfer to CH_4.

One of the more sophisticated techniques used in the study of the dynamics of ion–molecule reactions employs crossed beams of reactant ions and reactant neutrals, intersecting at a fixed angle of 90°. Ions from the collision zone pass through an energy analyser before going to a mass spectrometer, thus providing simultaneous measurement of the velocity and the angular distributions of the product ions. A schematic representation of this apparatus is shown in Figure 12.8.

The beam sources are mounted on a rotatable platform, which achieves the same effect as rotating the monitoring system with respect to the beams.

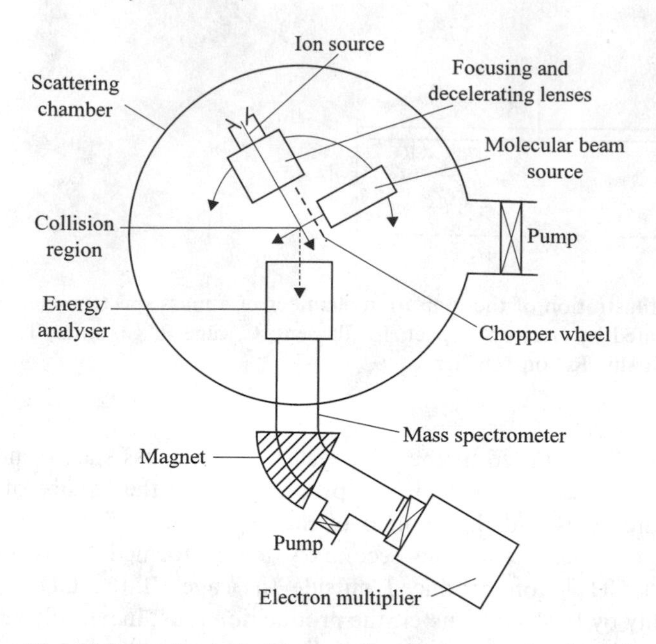

Fig. 12.8 A schematic diagram of an apparatus for the crossed-beam study of ion–molecule reactions. (Reproduced, with permission, from Herman, Z., J. Kerstetter, T. Rose and R. Wolfgang, *Faraday Soc. Disc.*, 1967, **44**, 123–36.)

The neutral beam is pulsed at a frequency of 200 Hz using a rotating chopper wheel, so that product ions formed in the reaction zone can be discriminated from those originating elsewhere.

For the reaction,

$$Ar^+ + H_2 \rightarrow ArH^+ + H^{\cdot} \tag{12.22}$$

with kinetic energies of the Ar^+ ions between 0.6 and 25 eV, the secondary ions were found to show a distribution in close agreement with that calculated on the model of spectator stripping. This model, or simple modifications of it, seems appropriate for a large range of ion–molecule reactions, especially at ion velocities well above thermal energies.

Another reaction studied by the crossed beams technique was the process,

$$C_2H_4^+ + C_2H_4 \longrightarrow C_3H_5^+ + CH_3 \searrow C_3H_3^+ + H_2 \tag{12.23}$$

for which it was found that, at ion energies of 1.4 or 3.3 eV, both the $C_3H_5^+$ and the $C_3H_3^+$ ions were distributed isotropically about the ± 90° axis. This indicates that the reaction proceeds through an intermediate complex

($C_4H_8^+$), with a lifetime of at least one rotational period. When this complex decays, only a little excess energy goes into translational energy of the fragments, so that the $C_3H_5^+$ ion is highly vibrationally excited and may dissociate by loss of H_2.

Experiments designed to measure the isotopic distribution of the products of the reaction of $C_2H_4^+$ with C_2D_4, or of $C_2D_4^+$ with C_2H_4, show that the relative amounts of the ions $C_3H_4D^+$, $C_3H_3D_2^+$, $C_3H_2D_3^+$ and $C_3HD_4^+$ are approximately 1 : 4.6 : 4.6 : 1. This is quite close to the statistical ratio of 1 : 6 : 6 : 1 which would be expected if, in the metastable parent ion ($C_4H_4D_4^+$) all the H and D atoms were equivalent and the three H or D atoms discarded in the methyl radical were selected randomly. Other experiments have measured the lifetime of the parent ion as around 3 ns, which is consistent with the extensive isotopic mixing that apparently occurs within this ion.

Suggested Reading

Levine, R. D. and R. B. Bernstein, 1987. *Molecular Reaction Dynamics and Chemical Reactivity*. New York: Oxford University Press.

Polanyi, J. C., 1967. *Discussions of the Faraday Society*, **44**, 293–307.

Lee, Y. T., 1987. Molecular beam studies of elementary chemical processes (Nobel lecture, 1986). *Chemica Scripta*, **27**, 215–28.

Polanyi, J. C., 1987. Some concepts in reaction dynamics (Nobel lecture, 1986). *Chemica Scripta*, **27**, 229–48.

Herschbach, D. R., 1987. Molecular dynamics of elementary chemical reactions (Nobel lecture, 1986). *Chemica Scripta*, **27**, 327–47.

Lias, S. G. and P. Ausloos, 1975. *Ion–Molecule Reactions: Their Role in Radiation Chemistry*. Washington, DC: American Chemical Society.

Ng, C.-Y., T. Baer and I. Powis, 1994. *Unimolecular and Bimolecular Ion–Molecule Reaction Dynamics*. London: Wiley.

Appendix A

Here we deal with the mathematics involved in deriving the RRK theory (Section 5.3). The first issue is the probability that, if the total vibrational energy of a molecule is ε in n modes, there will be at least ε^* in the critical vibrational mode. Let us assume that energy ε corresponds to j vibrational quanta and that the required number of quanta in the critical mode is m.

Assuming random distribution of these quanta throughout the n vibrational modes, this probability is given by:

$$\frac{P_{\mathrm{M}}}{P_{\mathrm{A}}} = \frac{\text{Number of complexions with } n \text{ in the critical mode}}{\text{Total number of complexions}}$$

$$= \frac{(j-m+n-1)!/(j-m)!(n-1)!}{(j+\mathrm{n}-1)!/j!(\mathrm{n}-1)!}$$

$$= \frac{(j-m+n-1)!\,j!}{(j+n-1)!\,(j-m)!} \tag{A.1}$$

It is reasonable to assume that j, the total number of vibrational quanta, is much greater than n, the total number of vibrational modes. Then we may say:

$$\frac{j!}{(j+n-1)!} = \frac{1}{(j+n-1)(j+n-2)\ldots(j+1)}$$

$$\approx \left(\frac{1}{j}\right)^{n-1} \tag{A.2}$$

If also $(j-m)$ is much greater than n, then we have:

$$\frac{(j-m+n-1)!}{(j-m)!} = \frac{(j-m+n-1)(j-m+n-2)\ldots(j-m+1)}{1}$$

$$\approx (j-m)^{n-1} \tag{A.3}$$

Thus we have

$$\frac{P_{\mathrm{M}}}{P_{\mathrm{A}}} = \left(\frac{j-m}{j}\right)^{n-1} = \left(\frac{\varepsilon-\varepsilon^*}{\varepsilon}\right)^{n-1} \tag{A.4}$$

Secondly, there is the integral (equation 6.15) which results from the product of the probability given in equation (A.4) and the Hinshelwood formula that a molecule will have a certain amount of energy, ε, in n vibrational modes (equation 6.14), namely:

$$I = \int_{\varepsilon^*}^{\infty} \nu \frac{1}{(n-1)!} \left(\frac{\varepsilon}{k_B T}\right)^{n-1} \left(\frac{\varepsilon - \varepsilon^*}{\varepsilon}\right)^{n-1} \mathrm{e}^{-\varepsilon/k_B T} \frac{\mathrm{d}\varepsilon}{kT} \qquad \text{(A.5)}$$

It is helpful here to make the substitutions,

$$x = (\varepsilon - \varepsilon^*)/k_B T$$

and $b = \varepsilon^*/k_B T$

Consequently, $\mathrm{d}x = \mathrm{d}\varepsilon/k_B T$. Thus, the integral becomes:

$$I = \nu \frac{1}{(n-1)!} \mathrm{e}^{-b} \int_0^{\infty} x^{n-1} \mathrm{e}^{-x} \, \mathrm{d}x \qquad \text{(A.5a)}$$

The definite integral in equation (A.5a) is known to mathematicians as the gamma function and, where n is an integer, this has the value:

$$\Gamma(n) = (n-1)! \qquad \text{(A.6)}$$

Thus we have,

$$I = \nu \mathrm{e}^{-b} = \nu \mathrm{e}^{-\varepsilon^*/k_B T} \qquad \text{(A.5b)}$$

which is the answer used in Section 6.3.

Appendix B

Solution of the simultaneous equations arising from the chain decomposition of H_2O_2, in the presence of copper(II) ion (Section 7.5).

It is convenient to use simpler terminology, as follows:

$$[H_2O_2] = x \qquad [Cu^{2+}] = y \qquad [OH] = u$$

$$[HO_2] = v \qquad [Cu^{+}] = w \qquad \phi I_a = k_0$$

The simultaneous equations in (7.55) then become:

$$2k_0 - k_1ux + k_6wx = 0 \tag{B.1}$$

$$k_1ux - k_5vy - k_7vw = 0 \tag{B.2}$$

$$k_5vy - k_6wx - k_7vw = 0 \tag{B.3}$$

By adding equations (B.1) and (B.2), we obtain an expression which does not contain u:

$$2k_0 - k_5vy + k_6wx - k_7vw = 0 \tag{B.4}$$

From equation (B.3) we have:

$$v = \frac{k_6wx}{k_5y - k_7w} \tag{B.5}$$

Substituting this into equation (B.4) yields an equation containing neither u nor v: it is thus an expression for w in terms of measurable concentrations and the various rate constants. However, it is a quadratic equation:

$$k_6k_7xw^2 + k_0k_7w - k_0k_5y = 0 \tag{B.6}$$

The solution to the equation,

$$az^2 + bz + c = 0$$

is well known to be

$$z = \frac{-b \pm \sqrt{b^2 - 4ac}}{2a}$$

$$= -\frac{b}{2a} \pm \left(\frac{b^2}{4a^2} - \frac{c}{a}\right)^{1/2}$$

In equation (B.6), the concentration w cannot be negative, so the negative root is discarded. Thus we have:

$$w = -\frac{k_0}{2k_6x} + \left(\frac{k_0^2}{4k_6^2x^2} + \frac{k_0k_5y}{k_6k_7x}\right)^{1/2} \tag{B.7}$$

From equation (B.1) we have:

$$k_1ux = 2k_0 + k_6wx$$

The net rate of consumption of H_2O_2 (equation 7.54) is:

$$-\frac{d[H_2O_2]}{dt} = k_0 + k_1ux + k_6wx - k_7vw$$

$$= 2k_0 + 2k_6wx$$

$$= 2k_0 + 2k_6x\left[\left(\frac{k_0^2}{4k_6x^2} + \frac{k_0k_5y}{k_6k_7x}\right)^{1/2} - \frac{k_0}{2k_6x}\right]$$

$$= k_0\left[1 + \frac{4k_5k_6xy}{k_0k_7}\right]^{1/2} \tag{B.8}$$

This equation differs only very slightly from that presented by Baxendale, J. H. and J. A. Wilson, *Trans Faraday Soc.*, 1957, **53**, 344–56.

Answers to the problems

Chapter 1 **1.1** $E_a = 208.4\,\text{kJ mol}^{-1}$; $A = 5.1 \times 10^{12}\,\text{s}^{-1}$. **1.2** Second order, $2k = 1.9 \times 10^8\,\text{dm}^3\,\text{mol}^{-1}\,\text{s}^{-1}$. **1.3** $k' = 101\,\text{s}^{-1}$. **1.4** $E_a = 262.4\,\text{kJ mol}^{-1}$; $A = 6.6 \times 10^{13}\,\text{s}^{-1}$. **1.5** $E_a = 83.1\,\text{kJ mol}^{-1}$; $A = 1.25 \times 10^{13}\,\text{dm}^3\,\text{mol}^{-1}\,\text{s}^{-1}$. At 70°C, $k = 2.75\,\text{dm}^3\,\text{mol}^{-1}\,\text{s}^{-1}$. **1.6** 95.0 days. **1.7** Order w.r.t. HBr = 1.0; order w.r.t. NO_2 = 1.0. Initial rate = $2.23\,\text{torr min}^{-1}$.

Chapter 2 **2.1** $k = 4.4 \times 10^{-3}\,\text{s}^{-1}$. **2.2** $k = 26\,\text{dm}^3\,\text{mol}^{-1}\,\text{s}^{-1}$. **2.4** $k = 2.2 \times 10^{-3}\,\text{s}^{-1}$. 2.5 $k = 2.02\,\text{dm}^3\,\text{mol}^{-1}\,\text{s}^{-1}$.

Chapter 3 **3.1** 2.0 with respect to NO_2, 1.0 w.r.t. N_2O_4.
3.2 $t_m = \ln(k_2/k_1)/(k_2 - k_1) = t_f/2$. **3.3** $k = 1.8 \times 10^4\,\text{dm}^3\,\text{mol}^{-1}\,\text{s}^{-1}$.
3.4 Decay is due to the reaction, $Ph_2\dot{C}OH + Ph_2CO^- \rightarrow$ dimer; $k_2 = 1.0 \times 10^9\,\text{dm}^3\,\text{mol}^{-1}\,\text{s}^{-1}$. **3.5** $k' = 1.6 \times 10^{-4}\,\text{s}^{-1}$.
3.6 $k_1 = 2.50 \times 10^{-3}\,\text{s}^{-1}$; $k_{-1} = 5.5 \times 10^{-4}\,\text{s}^{-1}$. **3.7** Both lead to $d[COCl_2]/dt = \chi_1[CO][Cl_2]^{3/2} - \chi_2[Cl_2]^{1/2}[COCl_2]$, but with different constants making up χ_1 and χ_2. They cannot be distinguished kinetically. **3.8** (a) 0.00597 g, (b) 0.0144 g.

Chapter 4 **4.1** 9.0×10^{18}. **4.2** $k = 1.4 \times 10^3\,\text{dm}^3\,\text{mol}^{-1}\,\text{s}^{-1}$.
4.3 $-95.7\,\text{J K}^{-1}\,\text{mol}^{-1}$. **4.4** $2.25 \times 10^{10}\,\text{dm}^3\,\text{mol}^{-1}\,\text{s}^{-1}$.
4.5 $6.5 \times 10^{10}\,\text{dm}^3\,\text{mol}^{-1}\,\text{s}^{-1}$. **4.6** 0.81 nm.

Chapter 5 **5.1** $14.15\,\text{kJ mol}^{-1}$; $-140.8\,\text{J K}^{-1}\,\text{mol}^{-1}$. **5.2** $-32.0\,\text{cm}^3\,\text{mol}^{-1}$.
5.3 $46.6\,\text{kJ mol}^{-1}$; $-102.7\,\text{J K}^{-1}\,\text{mol}^{-1}$. **5.4** Slope = +8.5.
5.5 (a) $z_A(z_B - z_C) = -1$ (and thus $z_A \neq 0$); (b) $z_A = -1$. **5.6** +1.14.
5.7 Yes.
5.8 If TS were symmetric, the k_H/k_D ratios for d_3 and d_4 isomers would be the product of the k_H/k_D values for d_1 isomers. These factors are 2.260 and 2.477, the experimental factors are 2.565 and 2.862. **5.9** $9.3 \times 10^7\,\text{dm}^3\,\text{mol}^{-1}\,\text{s}^{-1}$.

Chapter 6 **6.1** First order, $k = 7.9 \times 10^{-4}\,\text{s}^{-1}$. **6.2** No. **6.3** The ratio,

D_2 : D_2 : D_2 should be equal to 2 : 1 : 1.

Chapter 7 **7.1** (a) M + N = P + Q; (b) rate = $k_2(k_1/k_4)^{1/2}[M][I]^{1/2}$. **7.2** (a) Rate = $(k_1k_3k_4/k_5)^{1/2}[C_2H_6]$; (b) (iii). **7.4** (a) 7×10^6; (b) [Br], the same; [H], much greater. **7.5** Rate = $(k_1k_3/k_4)[H_2]^2$. NO_2 is involved both in the initiation and in the termination step.

7.6 (a) Rate $= \dfrac{k_3 k_4 (k_1/k_6)^{1/2}[\mathrm{A}]^{3/2}}{k_4 + k_{-3}[\mathrm{PhMe}]}$; (b) $E_a = E_3 + \frac{1}{2}(E_1 - E_6)$.

Chapter 8 **8.1** 2.94×10^{17}. **8.2** $(k^0_\phi - k_\phi)^{-1}$ is a linear function of $p_{O_2}{}^{-1/2}$.
8.3 Adsorption of 1-propanol is inhibited by that of H_2O.
8.4 $kKap_A.bp_B/(1 + ap_A + bp_B)^2$.

Chapter 9 **9.1** (a) $3k_2(k_1/k_5)^{1/2}[O_3][X_2]^{1/2}$; (b) Step 5 is mutual and not linear termination; (c) Yes. **9.2** $k_{obs} = k_1 k_2[Ag^+][Cr^{III}]/(k_{-1}[Co^{2+}] + k_2[Cr^{III}])$; $k_{obs}{}^{-1}$ is a linear function of $[Cr^{III}]^{-1}$;

$$Ag^+ + Co^{3+} \underset{-1}{\overset{1}{\rightleftarrows}} Ag^{2+} + Co^{2+}$$

$$Ag^{2+} + Cr^{III} \overset{2}{\rightarrow} Ag^+ + Cr^{IV}$$

9.3 $k_o = 1.9 \times 10^{-3}\,\mathrm{min^{-1}}$; $k_I = 8.5\,\mathrm{dm^3\,mol^{-1}\,min^{-1}}$.
9.4 $v_{max} = 5.5 \times 10^{-5}\,\mathrm{mol\,dm^{-3}\,s^{-1}}$; $K_M = 2.4 \times 10^{-3}\,\mathrm{mol\,dm^{-3}}$.
9.5 Yes. $K_I = 0.05\,\mathrm{mol\,dm^{-3}}$.

Chapter 10 **10.1** $\tau^{-1} = 4k_1 a + k_{-1}$. **10.2** $k_1 = 2.4 \times 10^8\,\mathrm{dm^3\,mol^{-1}\,s^{-1}}$; $k_{-1} = 4.9 \times 10^4\,\mathrm{s^{-1}}$. **10.3** $\tau^{-1} = k_1(a + b) + k_{-1}/(1 + K_2)$. **10.4** Step 2 is immeasurably fast. Slope $= k_1$, intercept $= k_{-1}/(1 + K_2)$.
10.5 $k = 2.3 \times 10^9\,\mathrm{dm^3\,mol^{-1}\,s^{-1}}$. **10.6** $\tau_1{}^{-1} = k_1 + k_{-1}$; $\tau_2{}^{-1} = k_{-1} + k_2/(1 + K_1^{-1})$.

Chapter 11 **11.1** $\tau_f = 51\,\mathrm{ns}$. **11.2** ϕ^{-1} is a linear function of [Q]. **11.3** 5.5.
11.4 $2k/\varepsilon = 2.5 \times 10^7\,\mathrm{cm\,s^{-1}}$. If diffusion controlled, $2k = 2.0 \times 10^{10}\,\mathrm{dm^3\,mol^{-1}\,s^{-1}}$ and $\varepsilon = 800\,\mathrm{dm^3\,mol^{-1}\,cm^{-1}}$. **11.5** 0.70.
11.6 $3.3 \times 10^{10}\,\mathrm{dm^3\,mol^{-1}\,s^{-1}}$.
11.7 $G(Fe^{3+}) = G_{OH} + G_H + 2G_{H_2O_2} + 2G_H k_1[O_2]/(k_1[O_2] + k_2[Fe^{2+}])$. Plot $\{G(Fe^{3+}) - 8.15\}^{-1}$ against $[Fe^{2+}]/[O_2]$.

Index

Index of Chemical Reactions